A Crisis like No Other: Understanding and Defeating Global Warming

Robert De Saro

Clinton Township, NJ
United States

A Crisis like No Other: Understanding and Defeating Global Warming

Author: Robert De Saro

ISBN (Online): 978-1-68108-961-4

ISBN (Print): 978-1-68108-962-1

ISBN (Paperback): 978-1-68108-963-8

What is the meaning of life?

There is none.

Deal with it.

But my two sons and their families come awfully damn close.

ENDORSEMENTS

A Crisis Like No Other is an excellent discourse on global warming topics that defines the many facets of the problem and covers its solutions as well as the politics involved. The read is a pleasure, with a daring and engaging style, making it useful to inform the reader on this pressing problem. This is an unusually well-written cross-over book that is suitable for the general reader as well as for college students in climate change, sustainability, and environmental science courses.

It covers a multitude of complex topics in an easily accessible manner, no matter the reader's technical background. For the reader's benefit, it includes examples, clever stories and anecdotes, and is packed with meticulously researched information as well as relying on the author's own expertise.

I highly recommend it.

—Carlos Romero, Director, Lehigh University Energy Research Center; and Fellow at Lehigh University

***

Bob De Saro's analysis of the situation along with suggested pathways to address an existential threat to our planet is refreshing, educational as well as an entertaining read.

A Crisis Like No Other covers all the topics a reader needs to thoroughly understand and take action on global warming. From the psychology of global warming denial and how to see through lies and fake news, to the science of global warming, to what we must do to solve it, and much more.

This alone would make it a valuable read but it is also written in a convincing and entertaining matter, filled with insight and wit. You will not be able to put this book down.

I recommend it for both the general reader and as a collegelevel book on environmental and sustainability science.

—Diran Apelian, Distinguished Professor, University of California, Irvine

What this Book is About

"*Hell hath no fury like a woman scorned,*" and Mother Nature has been scorned, misused, ignored, and insulted for the past 250 years. She is one ticked-off lady.

Why?

Because ours is a world getting hotter. Consider:

• The 1980s was the hottest decade on record.

• Until the 1990s surpassed it.

• And then the 2000s came along and beat the 1990s.

• 2010s? You guessed it. It took the crown.

• Now, our present decade of the 2020s is beating them all.

It's undeniable that global warming is occurring and we are responsible for it. And it is not good since global warming is causing rising sea levels, bigger storms, hotter temperatures, flooding, and drought. I could go on. All of which affect our well-being and that of our children and theirs, even more.

Still, we can emerge from global warming, maybe not unscathed, but mostly intact. To do so requires understanding what global warming is, how we can defeat it, and the bridges connecting the two.

A Crisis Like No Other: Understanding and Defeating Global Warming provides this understanding. It consists of four parts. The first part covers the psychology of global warming denial, how to defend ourselves against it, and how to convince others of global warming's grave harm. Part II describes what global warming is. The third part answers the question *What makes us so sure?* Finally, Part IV provides a road map to defeating it.

Part I. We Have Met the Enemy and He Is Us—Pogo

Facts and evidence are not always what they seem. The reason? We all have built-in biases and unswerving beliefs that filter and sometimes distort the truth. For example, why is a refs whistle a travesty when called against our team but deserved punishment when called against our opponent? It's simple. We see what we wish to see. So it is with global warming. Part I describes how we filter facts, driven chiefly by our subjective beliefs, which then push some of us to deny global warming despite the overwhelming evidence.

With that understanding in hand, part I goes on to discuss how to convince people to take action on global warming. It is surprisingly straightforward. We convince people of global warming's harm not by overwhelming them with facts but rather by appealing to what they already believe and understand to be common sense. People will listen if our arguments are crafted to fit their beliefs.

But to convince others of global warming and to understand it ourselves, we must first see through the lies and half-truths that come at us every day. Part I concludes by describing how we are built to fall for lies and how to defend ourselves from them. Uncovering the truth is not difficult if we have the proper tools. Ten easy-to-learn and logical techniques are given, along with examples.

Part II. The What's and How's of Global Warming

With part I as our foundation, we can now open our minds to understand what global warming is. Part II starts by describing how greenhouse gases are created from burning fuels and how they skew our planet's energy balance, leading to rising temperatures. The description is clear, accurate, and easy to understand. But global warming is not just about rising temperatures, as important as that is. It is also causing monster storms, droughts, floods, diseases, political instability, food shortages, mass extinctions, and increased violence—all of which are described in this book, with reasons given for each.

But to prioritize our actions, we need to know the source of the worst greenhouse gas generators. Mostly it's carbon dioxide (CO_2), but methane and some other bit actors play a role as well. The sources include industry, electric generating plants, and transportation. Oh, and cow burps.

Part III. Why We Believe Global Warming Is Real and Significant

Irrefutable evidence proves global warming is real and threatening. It's the same as when a thermometer placed under your tongue shows you have a fever. There is no one denying it. In the same way, with global warming, there are myriad measurements, mathematical models, experts from around the world, and much more, all converging onto this one unambiguous truth.

But to believe the evidence, we must also have a sound understanding of how science works, including its triumphs and failures. Science deals exclusively with facts, data, and measurements that either confirm our view of reality or reject it as being untrue. This process is repeated until our understanding of reality matches the evidence and is therefore confirmed. Part III describes more, but the evidence is science's bedrock and North Star.

However, nothing in life can ever be certain, including science. Part III explains the limits of how accurately we can know anything. Two types of uncertainties are explained: (1) that which is due to our limited but growing knowledge and (2) that which will forever be out of our reach since we are incapable of measuring (and therefore truly understanding) some parts of it. Part III goes on to explain how scientific decisions are made despite these uncertainties.

Still not convinced about global warming? Then ask those who have the most to lose. This part concludes by showing the military's grave concerns about global warming and its effect on their operations and bases. The military never kids around, and their anxiety about global warming should also be ours. Similarly, the insurance and financial sectors are getting sweaty palms over what could happen to their insured and investments.

Part IV. The Final Verdict

The last part of this book explains how bad the climate crisis is and what we must do to solve it. For instance, there are tipping points lurking out of sight, ready to pounce—ocean acidification, rising sea levels, melting glaciers, and thawing permafrost. But one tipping point that often gets overlooked is the ominous danger to our democracy. Global warming could lead to civil strife, creating opportunities for those wishing to sidestep our Constitution to acquire unlawful powers. We must stay alert to these dangers and add them to our incentives for fighting global warming.

Nevertheless, there is a clear path to solving our climate crisis. Three activities must be undertaken. The first is the technologies we need to deploy. The second is the decisions we must make in choosing our leaders and sidestepping the influencers who have fossil energy hidden agendas. And finally, it's all about us—the personal actions we need to take.

A Crisis Like No Other starts with the psychology of denial, moves to understand global warming and the science behind it, and wraps up with a road map on what we must do. That's how we beat global warming.

Mother Nature would approve.

A Few Additional Notes on the Book's Organization

Each part, indeed each chapter, can be read independently of the others, so you can skip around if you wish. For instance, if you want to get into the meat of what global warming is, then go straight to part II. Or if you are a direct action, no-nonsense sort of person, then go to part IV to get started on what we must do. Besides, you now own this book, so I suppose you can do as you please and never mind what I think.

Except I will need to walk that back a bit—not the owning thing, the jumping around thing. Some of the chapters refer to two or three ideas developed in chapter 1, so maybe it would be best to read that one first (after all, it's the first chapter for a reason).

For your enjoyment, I end each chapter with an "Afterthought," which is an engaging and sometimes humorous short take on one of the points in each chapter. But no peeking. Read the chapter first before the Afterthought. Yes, I know. You own the book and you'll do whatever you very well please. We already established that point. I'm just suggesting, is all.

I also sprinkle in "Asides" throughout the book, which provides interesting tidbits right after a particular concept is developed. You have no choice but to read them in the order I wrote them since I am not going to tell you where they are. I think I won this round. We're tied at one all.

Bye for now. I'll see you in chapter 1. Or whichever chapter your fancy takes you to.

CONTENTS

Part I. We Have Met the Enemy and He Is Us—Pogo

Part I covers the psychology of global warming denial, how to convince others of the crisis, and how to see through the many lies and conspiracy theories that can easily confuse us.

Why We Deny Global Warming

"There is absolute proof."

"They won't listen. You know why? Because they have certain fixed notions. . . . Any change would be blasphemy in their eyes, even if it were the truth. They don't want the truth. They want their traditions."

"We could try."

"We would fail."

—Isaac Asimov, Pebble in the Sky, 1950

Global warming is happening. It's happening now. It will continue to happen. And we are responsible for it. These facts are inescapable, and denying them—especially when you consider all that is at stake—is irrational to the point of madness. But deny it people do:

"The science is not settled, and the science is actually going the other way. . . . We may in fact be going into a cooling period."

—Joe Barton, Former Congressman (R-Texas)

"Climate scientists deserve to be flogged."

—Marc Morano, ClimateDepot.com

"On average, global warming is not going to harm the developing world."

—Bjorn Lomborg, Copenhagen Consensus Center

"The Great Hoax: How the Global Warming Conspiracy Threatens Your Future."

—Senator James Inhofe (R-Oklahoma)

"CO_2 plays only a minimal role."

"More CO_2 will benefit the World. The only way to limit CO_2 would be to stop using fossil fuels, which I think would be a profoundly immoral and irrational policy."

—Michael Halpern, Program Manager, Union of Concerned Citizens

Despite these wacky statements, people rarely make illogical decisions. As a species, we have flourished for over two hundred thousand years. Not as good as the dinosaurs, certainly, but not bad at all. Do you think we would still be around if we made reckless decisions not driven by our survival needs? We would not. Throughout our brief history we have always made sound choices, and, outward appearances aside, those choices have been steeped in logic and need. The above statements are no exception.

Then what gives?

The answer involves understanding how we think, how we make decisions, what subconscious forces drive our decisions, and how they help us survive and thrive. Most importantly, it's about how we mercilessly discard facts that don't support our beliefs.

CHAPTER CONTENTS

Quick Assessments are Critical for Our Survival

The alien spaceship lands silently and without notice as the planet's inhabitants bustle about their lives like ants around a crumb of food. From the interior of the metallic ship emerges a robot bristling with sensors for anything it might encounter. But on this run, it has just a single and simple mission: to determine how the denizens of this backwater planet behave.

It surreptitiously records every action people make, including the glucose their brains consume, the decisions they make, their reactions to events, and their brains' computing power.

After several days it reports to its superiors and it is perplexed. "How can this be? These bimodal creatures walk to their cars, drive them, avoid traffic, play sports, all smoothly and quickly with no conscious thought going into it. And when I measure the energy required by their brains for such activities, it is much too small. It is simply not possible to go through all those calculations with such little energy use. Even these primitive carbon-based life forms know science prohibits such things. And their tiny craniums house a brain too small to accomplish much beyond the minimal chores needed to survive. Yet just the other day, a driver was approaching a truck crossing its path, and he avoided it as deftly as a supercomputer on steroids. It should have taken 320 trillion calculations per second to accomplish this, well beyond the spare computing capacity of these hairless apes. There is no possible way they can be doing this.

"But it gets worse. I challenged them with a series of problems. I watched as they tried to determine the rate of return on an investment. And guess what. Their behavior altered so dramatically that I thought I was examining an altogether different species. Their brain glucose consumption shot up. They did their tasks with intense focus, exertion, and determination, unlike their previous efforts. Plus, they were no longer vigilant and could be easily spooked. I had a prankster pop a paper bag in front of these so-called sapiens while they were deciding on a movie to rent, and they jumped like one of their planet's marsupials. I would have laughed if I were capable of humor.

"It is as if there are two beings in the one body. One that glides through life as if on a magic carpet with little effort, and the other takes over when faced with a new problem not in its database. And it is seamless. In fact, I am sure these organic creatures don't suspect the dual nature of their essence—the two beings living inside the one body."

Our robot interloper is correct, and its unease at us having two beings in a single shell is close to the mark. Daniel Kahneman [1] and Amos Tversky, in a series of clever and amusing experiments, uncovered our dual nature, calling it System 1 when we are on autopilot and System 2 when we are deep in thought. And it is these two intrinsic human qualities that contribute to our global warming confusion.

Just as the extraterrestrial robot had discovered, System 1 is continuously vigilant, always assessing our surroundings instantly and ready to act without hesitation. It is fast, as it must be, to protect us from life's daily assaults. It cannot have doubt or second guess itself—doubt takes time it does not have. It sees, it assesses, it acts. There is no other possible path for it.

System 1 provides a means to assess any situation quickly with little effort [2], which we use every waking moment. Try walking along a busy street and try thinking about all that is happening around you. You'd never make it; too much information. Instead, you assess the path in front of you and navigate it mindlessly and accurately tossing away extraneous details not needed to achieve your target.

Think back to a time when you lost your temper. Did you think about what your next step would be? Or did you simply give in to your emotions and either lash out or smolder in frustration?

It is the same way driving a car, going through our morning rituals, talking to a friend, and even sitting down for a meal. We don't think about what we are doing and we go about it effortlessly, almost unconsciously.

System 1 has some tricks up its sleeve to help its host dispatch all of life's bothersome intrusions. Here are a few.

System 1 is never stumped, even when there are times it should be. Sort of a friend who can't stop talking. They should, but simply can't help themselves. Similarly, instead of throwing in the towel and admitting its inability to assess a situation, System 1 just finds a way around the impediment. If a question posed that it cannot answer, it changes the rules of the game and answers a different question in its place. System 1 is nothing if not clever.

Here are some difficult questions with their possible substitutions that make life so easy for us.

Difficult Question	Substituted Question
Is global warming happening?	Was it hot today?
Is the professor a competent lecturer?	Am I being entertained by her?
Should I vote for Sarah?	Does she make me feel comfortable?

(Table) cont.....

Difficult Question	Substituted Question
How happy have I been this past year?	How happy am I now?
Should I approach this stranger?	Is the stranger good-looking and well dressed?
Should I buy This Car?	Can I visualize myself having a great time in it?
Do I like the Beatles?	Do my friends like the Beatles?
Did the ref make a bad call?	Was it against my team?

Kahneman reported on students being asked how happy they were [3]. But if that question was preceded by "How many dates did you have last month?" the results were much different, with Kahneman interpreting this as the students using the first question as a substitute for the second. They substituted the much easier question about their dates for the more difficult one about how happy they were. Most importantly, they didn't realize they had done so.

We cheat when we can. When we are asked to evaluate an argument or position, we often use peripheral clues in place of true content because it is easier to do. For instance, having a reasonably accurate ability to quickly determine if a stranger poses a threat goes a long way in our survival. Tzvetan Todorov showed that we use only two physical clues to judge if a stranger is safe or not: a person's chin and whether they are smiling. A square chin, denoting dominance, and a frown trigger our instincts to be wary. We can, at a glance, judge two essentially crucial facts: (1) how dominant a person is and therefore whether they are a threat or not, and (2) how trustworthy they are and therefore whether their intentions are more likely to be friendly or hostile. Even if we are not perfect at picking up the clues, any level of competence gives us a survival advantage and we are more likely to pass on our genes, which means our offspring will inherit the same traits.

But here's the thing. Todorov also showed that voters use that same mental tool to quickly decide if they will vote for a candidate. Of course, the voters' intuitive feel has no bearing on the candidate's abilities; rather it is just an easy way to answer a difficult and maybe an impossible question as to whom to vote for. A great shortcut most of the time, but in this example, it fails us.

In a similar vein, Daniel T. Willingham showed that instead of withholding judgment, we evaluate a speaker on their looks, clothing, mannerisms, accent, and other outward appearances [4]. Even though the speaker can twist the results or cite only favorable studies and ignore the ones that refute their position, we still may agree with their position if we identify with their looks. The more they are like us, the more we will agree with them irrespective of what the truth actually is.

As Noel E. Sharkey states, "[System 1] . . . overlooks contradictory information, neglects ambiguity, suppresses doubt, ignores the absence of confirmatory evidence, invents causes and intentions, and conforms to expectations" [5]. All to get us through another day.

System 1 is very efficient with our time and effort. And we can arrive at seemingly intuitive opinions on very complex issues in an instant. If System 1 is right most of the time, and if when it is wrong, it does so only in situations that are not critical to our survival, then it is doing a good job. On balance, it provides a valuable benefit.

But what happens when we must figure something out, something truly complex, and we have the time and security to do so? There is no danger, so System 1's emergency reactions are not needed. Further, we can't use System 1 to unravel a complex ambiguous situation, as it cannot tolerate doubt and certainly can't weigh different ideas or choices.

Enter System 2, the analytical and deep thinking part of our being.

System 2 is deliberate, is in no hurry, and uses the full logical and mental skills available to us. For instance, try this simple exercise. Rearrange the following words in alphabetical order.

- Zebra
- Cat
- Telephone
- Bath
- Math
- Book
- Soldier
- Mickey
- Compassion

You could not do it by inspection (System 1), so you called upon System 2 to look at the first letter of each word and compare it to another and iterated until the list was in the correct order.

Here's another example. Look at the animals in the figure below, and then answer the two questions that follow:

1. What two animals are pictured here?

2. Are there more cats than dogs?

Just visually inspecting the figure with no conscious thought was all you needed to answer the first question. Your System 1 was engaged. Not so with the second. Here you needed to count the cats and the dogs and then compare each. It took effort and focus. If someone interrupted you, you would need to start all over again. That was System 2 at work.

One more example. Answer this question as quickly as you can:

You are running in a race when you pass the second-place runner. What place are you in?

Now answer that same question, but instead of jumping to an answer, draw three stick figures representing three runners moving from left to right on your paper. Label the last one with your name. Next, draw a line from your position to just in front of the second-place runner showing you passing her. Now, what place are you in?

Chances are, the first way you answered this question you used your System 1 and your answer was first place. The second time, using System 2, you got the correct answer, second place. System 1 fooled you, and though it may have been close enough for most purposes, for this one it would have been wrong.

System 2 has significant advantages over System 1. It can accept doubt, and it can weigh two or more opposing concepts for consideration. Essentially, it can think critically and examine issues that would be impossible for System 1.

And it can overrule System 1's impulsive actions. This is the source of our inhibitions and better decision-making. This is where we exercise constraint and willpower. System 1 will go for the chocolate cake, yearning for instant gratification and calories. System 2 will put the brakes on our impulse, willing to trade instant gratification for long-term health.

System 1 is emotional whereas System 2 uses logic. System 1 is fast, nonverbal, and motivates us effortlessly based on our feelings and without our awareness. System 2 is slow and effortful and is where our willpower resides.

We fall in love using System 1. We pass our math and physics exams with System 2.

But System 2 comes with a boatload of disadvantages. Unlike System 1, which can multitask, System 2 can do only one thing at a time. We can perform multiple tasks simultaneously only if they don't require deep thought. The first time we learn something new requires all our attention blocking out any other activities. For instance, the first time we tried driving with a stick shift, the car bucked and stalled while we concentrated solely on it. But once it became second nature, we downshifted, listened to the radio, weaved through traffic, all the while drinking our coffee with little effort.

System 2 requires much greater effort and expends much more energy and glucose than does System 1 [6]. That is why thinking is so hard and why we evolved System 1 to take the load off. In our early world, food was scarce, so limiting System 2's workload helped us survive.

David DeSteno mentioned another problem using System 2 [7]. Teenage students using System 2 for better self-control had more success but suffered from increased stress and had premature aging of their immune cells. Wow! This seems a bit much, and suggests we should just abandon our willpower and traipse through life unfettered by responsibilities or social obligations—just live for the moment from day to day. I'm not fully buying into this, but perhaps it raises a more nuanced issue: that the cost of System 2—even if not as dramatic as reported here—still has an effect on us, all the more reason to have System 1 take over whenever it can and when careful deliberation isn't called for.

System 1 doesn't operate in our conscious minds; it doesn't say, "Hey, Joe, turn the wheel now or you're going to become a pepperoni road pizza." It works instead by controlling our actions through our brain's hard wiring and by releasing chemicals into our bodies that force us to act.

System 1 resides in the amygdala part of the brain and intercepts information from the outside world before it gets to the cerebral cortex and into our conscious minds [8]. It hides out and hijacks the information while we remain clueless. So much for being the master of our fate.

System 2, on the other hand, is centered on the prefrontal cortex (PFC) of the brain, which is critical for executive-type decisions. The PFC is the most advanced part of the brain and is what makes us human. It handles higher-order cognitive abilities, is crucial for inhibiting inappropriate behavior, and fosters creativity.

But Quick Assessments Can Fool Us

Despite System 1's essential attributes, it can also fool us. When we rely on System 1, we set ourselves up to make quick decisions using just the information at hand with no thought or analysis. We give up critical thinking and we fail to judge the value of the information, instead accepting it unconditionally. And we are impulsive, impatient, and rely on the first answer that comes our way. System 1 prefers the simple to the complex; simple is much easier to process and the complex requires work that we would prefer not to do. System 1 is the teenager in us all.

A large number of experiments clearly demonstrate how System 1 is not always our friend. And taking this a step further, System 1 not only acts without our awareness, but sometimes against our will. Scary but true. Let's see how.

Kahneman describes an experiment carried out at New York University in which a group of students were told to unscramble a series of five words into coherent sentences. One group had words associated with the elderly (*e.g., Florida, forgetful, bald*) while the control group had no such words. When asked to walk across the hall, the students primed with the elderly words walked significantly slower than the control group, mimicking the elderly. Further, when questioned later, none of the primed students were aware that the words had an elderly theme and none realized they had walked slowly. However trivial the effect was, they nonetheless were manipulated into acting a certain way and they complied with no awareness they were doing so.

In a similar experiment, students were primed with money words. The primed students were more independent than the control group but were much more selfish as they were less inclined to help a student pretending to be stumped on a different experimental task. Again, the students were unaware their behavior had been intentionally changed.

Kahneman also notes a more insidious use in "Dear Leader" pictures on walls everywhere which primes for obedience and inhibits independent thought. Again, people are manipulated—in this case into obedience—without their knowledge and certainly without their consent.

Our Metabeliefs Organize Everything we Know and Care About

We know the sun rises in the East, where we need to go each morning when we leave the house, where we will stop for our morning coffee, which is our favorite baseball team, who our friends are, that tigers bite and rocks don't. And as the alien robot noted, we do so as if on a magic carpet, without conscious thought.

Just as we could not survive very long without Systems 1 and 2 operating in their respective kingdoms, we could not survive without our Metabeliefs intact and driving our decisions and actions. How could it be otherwise? It would not be possible to reload all these facts every time we needed them.

Metabeliefs are our GPS in navigating our lives. They are the sum of our knowledge, feelings, emotions, facts, intuitions, instincts, cultural norms, and genetics that constitute the whole of what we understand and believe. We acquire our Metabeliefs throughout life as we experience and evaluate whatever we sense and feel from our environment, friends, relatives, and whatever else impacts us. Mostly we are unaware of this enormous infiltration that shapes who we are and how we act. But not always. We can change our Metabeliefs by letting System 2 critically think about something and absorbing its conclusions into our being. But mostly it is System 1 sculpting our Metabeliefs without our awareness of the transformation taking place over a lifetime. I suspect that in our early years our Metabeliefs are rapidly formed, whereas as we age, we add less and less to our accumulated beliefs.

Asking our brain to process the enormous information available to us is impossible so it must take shortcuts. Our mind filters whatever it sees and what filter can it use? The only one available to it: our Metabeliefs, the mother of all filters. Consequently, what we see, hear, and feel will always be in concert with our Metabeliefs, for good or bad.

And our Metabeliefs must construct a cohesive story, using the available facts, about what is going on so it can take action when needed. And it must do so in milliseconds. And then repeat again, again, and again. The story must be consistent with our own beliefs. It would never provide a scene that was in conflict with our beliefs and our understanding of how the world works and how those near us should behave. It cannot act differently.

The huge benefit of our Metabeliefs is that we can take a very small fraction of what we see and hear and our brains will process that small sample and make a whole picture, filling in what we need and ignoring what is irrelevant to our survival. Anne-Claude Gingras estimates that we can process only 0.0000005 percent of the information our eyes and ears receive [9]. That's like watching a two-hour movie but seeing only 0.004 milliseconds of it. Really tough to get the gist of the plot, but somehow we manage.

Do this experiment. Go somewhere that is visually rich—say, a mall. Slowly scan the stores, the people, the food court, products, colors, the floor, people wearing different clothing, the smells—nearly an infinite amount of detail if you look carefully. Now pick a store and walk toward it with that being your only goal. You will register only the fewest of details, just enough to get by. No more, no less. What happened to those other details that are clearly there? Your mind filtered them out; otherwise it would have blocked your main goal of getting to that specific store.

So it is not so much what we see but what our mind thinks we are seeing by sampling a small part of the external world and quickly and efficiently creating a composite picture. One rule it uses to great effect is that what it already knows about the world must be satisfied by its new construct. If it isn't, then the brain alters what we see until the view conforms to what is sensible to our Metabelief. It's imperfect, to be sure, but there is no other way if we are to survive from moment to moment since there is far too much information for us to handle. Players immersed in combat video games, such as *Halo*, learn this fact quickly. Taking their eyes and focus off the attacking aliens for just a moment, and the player becomes toast.

"As an Aside, this is why eyewitnesses are not always reliable. Most events occur so quickly that the eyewitness has only a brief time to see it, which is a perfect setup for System 1 and Metabeliefs to take over. Unfortunately, this is not an accurate way to assess what is happening when details become crucial. For instance, since 1989, eyewitness testimony that would have resulted in the death penalty for 254 people was overturned by DNA evidence [10]".

But Our Metabeliefs will Fool us as Well

But there is a rub. For our Metabeliefs to operate quickly and efficiently, they must sometimes falsely manipulate information the way a crooked tax accountant will claim our dog as a dependent. It does so innocently enough, since dawdling over what appears to it as unimportant details will derail its efforts to keep us physically and mentally whole. It is simply a shortcut that evolved over many years, and it works remarkably well.

Mostly our Metabeliefs are correct. Mostly.

Road rage is an example of Metabeliefs gone nuclear. A perceived slight from another driver leads to an aggressive counter, not because there was a physical threat but because the offended person feels his beliefs about who he is are being challenged. His ancient instincts about his social status inappropriately kick in, as the slight would threaten his group standing and therefore his ability to acquire food and mates, which demands a quick and violent response. It overreacts because it knows no other way.

Here are some examples of how Metabeliefs can affect our opinions on global warming:

• People's understanding of religion is a Metabelief that can make it difficult to believe in evolution or global warming.[i] To do so would put their religious beliefs in jeopardy. "Nature is God's exclusive handiwork and it is arrogant to believe we can upset His work." And there is the belief that God has given humanity dominion over the earth, to exploit as we wish.

• Global warming contradicts America's heroic image of itself and the belief that we will inexorably become more prosperous. The fear that global warming will arrest our nation's progress and show that America is not omnipotent is too difficult for some to accept.

• The belief that the world is just and innocent people do not deserve to suffer due to global warming. I fell victim to this one. There was a period in my life when I

---

[i] Major religions accept global warming and our role in it, so it is not religion itself that's the problem but our skewed and self-interested interpretation of it.

stopped reviewing the global warming literature because it was too depressing.

• Some people resist authority, especially academic or intellectual authority.

• Global warming resistance is often a proxy for the real issue—we don't want government interference or additional costs. This is especially true for people who feel they have been marginalized and have lost economic ground compared to their expectations.

• California farmers are worried more about climate policy than climate change [11]. They are not skeptics of global warming but rather skeptics of regulations that will affect their livelihood: "We can adapt to climate change but not regulations." They don't care if the world ends in one hundred years if regulations effectively end their world in the next one.

People's Metabeliefs include their political affiliations to such an extent that their membership becomes more important than what the group stands for [12]. Republicans are proud they are not Democrats just as Democrats are proud they are not Republicans. This pushes them to accept the beliefs of their group without having to evaluate its truth. So merely being a Republican pushes us to deny global warming whereas being a Democrat pushes us to accept it.

Psychologist Asheley R. Landrum states, "People with more knowledge only accept science when it doesn't conflict with their preexisting beliefs and values" [13]. Landrum produced experimental evidence for this effect. She had people read articles that linked the disease Zika to either global warming or immigration. When connected to global warming, there was an increase in concern about Zika from Democrats and a decrease from Republicans. And the opposite occurred when it was linked to immigration. People's beliefs have less to do with what they know and more with who they believe they are.

The fact is, global warming denial is not about being uninformed [14]. Global warming issues are getting more press than the Kardashians so you would need to be living under a rock to miss it. No, it is about what our Metabeliefs choose to do with the information.

Equally Importantly, The Groups We Belong To Help Shape Our Metabeliefs

Would you rather walk down a dark alley in a strange city alone or with a group of friends? Easy choice. Survival and good fortune depend on being part of a group and sticking together. A group provides the cohesion and the shared beliefs that lead to coordinated action, which not only increases the survival of any individual but also allows the individuals to flourish beyond what they could do on their own. And the stronger the cohesion, the more effective the group and the more benefits to its members. Of course, we can go too far, which allows charismatic leaders with bad intentions to control us, as evidenced by the Jonestown Massacre [15]. But on the whole, if the group has the right balance of adhesion and commitment, it will do well.

Groups provide much more than physical protection. They also affirm our identity and how we view ourselves. They provide us with values that are reinforced by other members. They prescribe how we should behave, think, and how to make sense of the world [16]. They remove uncertainty in how we go through life and they organize us to achieve much. Imagine trying to build the pyramids or land an astronaut on the moon without deep group adhesion and member acceptance.

But it comes with a price. We lessen our independent thinking and replace it with groupthink. We become less critical of our positions and suspicious of other ideas since they represent an attack on our group that jeopardizes the benefits we receive. Maybe we don't consciously think of it this way, but our System 1 is certainly pushing our actions in that direction. In fact, experiments have shown that when people conform to group ideas, their prefrontal cortex is suppressed, which is where System 2 resides.

So when global warming statements threaten to upend a group's beliefs, they get tossed into the waste bin because they attack what are most important to the group's members. Sure, you can argue that global warming will hinder our breeding, at least in the long term, so shouldn't that be just as important to the group? No. First, people react to immediate threats, not future ones, especially when their needs for food and sex are here and now. Second, global warming is everyone's problem, so let someone else take care of it: "I'll hold on to my current gains and let others handle the long-term stuff. Better they pay the price than me." Third, "global warming cannot be real if it conflicts with my beliefs and it disadvantages my group and therefore what I gain from group membership. After all, it does me no good to fight global warming if it destroys my group and I am left with nothing. I'd rather die than face that."

Further, we are social learners [17]. Our beliefs come from those we trust: teachers, friends, relatives, and people we consider part of our group. This often works well, but not always. For instance, it fails miserably when anti-vaxxers trust evidence from their friends and others in their orbit more than evidence from the Centers for Disease Control and Prevention or other impartial medical organizations. An amplifying problem is that we are also conformists. We wish to get along with our group members and avoid rocking the boat. In most matters, such as outside threats, that is the only way to survive. But when the group gets it wrong, it leads to poor choices with consequences beyond the group. When information challenges one's Metabeliefs, and therefore the group itself, people are forced to "vigorously defend their values, identities, and attitude at the expense of factual accuracy" [18].

Our Metabeliefs also compel us to look upon other groups as deficient or evil because they are different from ours, which threatens our self-identity and well-being. If it is different, it must be feared and therefore hated.

Let's do a thought experiment.[ii] Build a red robot with superb artificial intelligence capable of quickly learning any skills you care to teach it. Put it in the woods and teach it how to hunt game, live off the land, protect itself from predators, interact with similar-looking red robots, and defend its group from other bands of robots. It doesn't always go smoothly. Occasionally the robot gets trashed by a wild animal, so you bring it back to your lab, repair it, and change the software part that didn't work but retain the rest. You then put it back into the wild, now that it can better deal with that particular danger. And this is a continual process. As time goes by, the robot gets better at surviving and thriving and you need to fix it less and less often. Sure, occasionally a random event happens—a tree branch falls on it or a large rock from space smashes into it, but by and large it does well enough to keep on ticking. It took a while and you had to fix a lot of mistakes, but it finally reaches a point of competence in the environment you placed it in.

Now comes the challenge. Take the red robot, put a tie on it, and turn it loose in an urban city with a suite of new challenges. The robot is confronted with problems different from those you painstakingly trained it for over so many years. It will put problems it encounters into imperfectly fitting pigeonholes according to what it previously learned. Its behavior will appear erratic and illogical at times but only because it is using software developed for a different environment.

It will encounter green robots that are clearly different from it and that have different ideas about how the world works. So what will our red robot do? The only thing it can: it will fall back on its training, what it knows and what has worked over so many years, and it will defend red robots at the expense of the green ones. It will distrust green robots, dislike them, and maybe attack them, even if such actions are not called for. Given what the red robot knows, how can we possibly expect it to act any differently?

We are not much different from the red robot, as our survival instincts and ways of dealing with problems were developed in an environment much different from the high-tech world we now inhabit.

[ii] Invented by Hans Christian Ørsted, thought experiments exist only in our minds but can provide surprising insight into difficult problems.

Don't believe me? Then try this experiment. Pick a group of friends who share a common bond with you—say, a sports team. Do something contrary, such as wearing a jersey from a competing team. What happens? It will certainly be noticed and you may be ribbed over it, in good nature I am sure, but there will be undercurrents of hostility simmering just out of sight. What have you done? You have gone against your group, and by doing so you have threatened their values, at least on this topic.

A young woman I know does not believe global warming is occurring, not because she carefully studied the issue but rather because her friends and family don't believe it. In her mind it would be a waste of time to investigate something her group already has, and since they have been correct so many times in the past, there is no reason to believe they are wrong now. Her group is tight-knit, and to go against the wisdom of the group would expend social capital she would rather save. To have a dissenter within the group would stress it, which could lead to damages greater than the single issue of global warming. For instance, dinners might become less fun, and trading business contacts might become less frequent or might end altogether. The social and economic penalties could easily outweigh this single issue, so she takes the sensible path and falls in step with her elders and peers. Many of us would.

My young acquaintance cannot articulate or consciously think of it in this fashion, nor is she even aware of the decision she is making. No need to. Her Metabeliefs do it for her.

Confirmational Bias Aids in the Subterfuge

How do we protect our Metabeliefs from factual assaults that don't jive with its core foundation? We eliminate the offending fact the way a dictator eliminates his opposition. Not surprisingly, System 1 is ideal for such a role. It is forever lurking beyond our vision ready to pounce on anything that makes us uncomfortable by either changing what it is or deleting it altogether. And by so doing, it keeps System 2 totally in the dark about what just occurred. There are two related techniques System 1 employs: *cognitive dissonance* and *confirmational bias*.

Cognitive dissonance is the setup man. It is the name given to our inability to hold two opposing ideas at once [19], especially when one of them conflicts with our Metabeliefs. For instance, when people drive gas guzzlers knowing it contributes to global warming, they are in a state of cognitive dissonance due to the two conflicting feelings they have: the first one is they enjoy driving big SUVs, and the second is they are worried about how global warming will affect their children's future. This produces a mental conflict that makes them uncomfortable and no one likes that. And since System 1 seeks always to be right—it has no time for ambiguity—it will interpret facts to fall within our Metabeliefs. Don't blame System 1. It's just doing its job.

Next, confirmational bias takes over to close out the problem. Confirmational bias stops information reaching us that opposes our Metabeliefs and warmly welcomes facts that support our beliefs. Our SUV driver above might use this tool any number of ways. She might limit reading about global warming, or eagerly embrace sales brochures that tout the improved gas mileage of SUVs, or decide global warming really isn't all that bad. There are a multitude of paths her System 1 can take to release the tension.

A sad example of this is a news story about a mom who took her young daughter to a quack doctor. The young girl died in his care, and as a result the doctor was prosecuted for malpractice. The mom vigorously defended the doctor, claiming he did no wrong despite substantial evidence to the contrary, which she refused to see. Why would she defend the doctor? Because she couldn't accept that she may have contributed to her daughter's death by taking her to him. To preserve her sanity, she had no choice. Confirmational bias was doing its job of protecting her.

Other examples of ignoring evidence: smokers rationalize that only heavy smokers get lung disease. Shopaholics rationalize that they are getting good deals. Anti-vaxxers rationalize that they are wresting control back from the government or that their religious beliefs are in peril. Some will view hot spells as confirming global warming is happening while others view cold spells as confirming the opposite. All of which lead to bad outcomes.

Commitment bias is a form of confirmational bias that we have all experienced. It occurs when we are unwilling to retract a position we have committed to, even when there is clear contrary evidence [20]. Backing off a publicly stated position displays weakness and would take us down a notch in our perceived social pecking order. That is why it is never a good idea to put an opponent in a corner. Unless you are capable and intend to annihilate them, it is best to give them a face-saving way of changing their position.

Mind you, we possess confirmational bias and cognitive dissonance because they help us survive; otherwise they would have been evolved out of us. And they are a very efficient way to operate since life is littered with inconsistencies and unexplainable stuff. Usually, these factoid blips are of no consequence and it would be a waste of time and glucose to try to figure them out. And while we were, a predator might just sneak up from behind, putting an unfortunate end to our deliberations. They also help System 1 move us through life unimpeded and protect our Metabeliefs from unpleasant and contrary facts as well as promote harmony in our groups.

So it is really important that we have confirmational bias working for us when we need it.

But *no* asset can ever be perfect and errors will always occur as we use these tools. Further, as with our red robot thought experiment, we evolved these traits in a world much different from the one we now live in, so it is not surprising that errors occur. And to add fuel to the fire, according to Sara E. Gorman [21], when we confirm facts as true, we get a dopamine rush similar to eating chocolate, having sex, or falling in love. Who knew. Though I suspect the effect is quite a bit less.

"As an Aside, confirmational bias is how stereotypes are created. We remember aspects that reinforce the stereotype but forget or don't see ones that oppose it. For instance, a common stereotype is that athletes are inarticulate. We see the ones who stumble when interviewed on TV, ignoring, of course, the stress and fatigue they are under having just completed a tortuous 3½ hour game. Yet it doesn't register that some of the most articulate sports announcers—Dan Dierdorf, Tony Romo, Chris Collinsworth, and Brandi Chastain, for instance—were once athletes".

Both liberals and conservatives are equally prone to confirmational bias, but on different topics. A common theme is that liberals tend to promote government regulations, regardless of the science quality, whereas conservatives (and libertarians) reject what they believe is liberal agenda-driven science and tend to oppose regulations even when the science is sound [22].

Liberals and conservatives also have their own global warming denial paths. Conservatives deny that global warming is happening or that we are responsible for it. They do so because they believe global warming actions will change their way of life, which they view as sacred, an unthinkable situation [23]. Liberals accept global warming facts but live in a fantasy world about how easy it will be to solve it. In this sense, conservatives come closer to the truth than liberals since they understand how difficult global warming solutions will be, but both bury their heads in the sand, leading to the same bleak outlook.

Now add a sprinkling of science distrust and voila—you have discredence, a willingness to disbelieve. And to make it long-lasting, add social acceptance and the echo chamber effect, in which one specific idea is repeated so often that it is assumed to be true just by virtue of the repetition. Mix and stir using Metabeliefs, System 1, cognitive dissonance, confirmation bias, and discredence and you have a fine cocktail potent enough to derail the truth of global warming by anyone so inclined.

Why the Poor Remain Poor

Our System 2 burns through glucose, so what happens when there isn't enough or when it gets fatigued? We unconsciously default to System 1 to make our decisions, which is not always the best option. A worrisome example of this is a study that followed Israeli parole judges and their decisions before and after meals [24]. Right after a meal they approved 65 percent of the applicants. But then their approval steadily dropped, reaching zero just before their next meal. Their decisions had less to do with the merits of each applicant and more to do with the judges themselves and how fatigued and hungry they were. Immediately before a meal, their System 2 was exhausted, so their System 1 automatically took over and went with the fast, easy decision to deny parole. This was not a conscious act. They were not deliberately trying to sabotage the judiciary. They were just victims of their genes driving their System 1 to act.

"As an aside, this is one reason to take your clients out to dinner. If you want a favorable decision, be sure your clients are not hungry and are in a good mood so they will not default to the easy answer of no. As another Aside within an Aside, this is also a sales technique known as the reciprocity theory, in which the client feels the need to reciprocate for the free meal".

Mental exhaustion also works against the poor which helps explain why it is difficult to pull oneself up by the bootstraps and out of poverty. Self-control is critical in achieving goals such as improving one's finances. Because the poor have less money and other resources, and are often overwhelmed with day-to-day problems, they need to restrain themselves more often than wealthy people. Thus, their willpower becomes depleted and they are less able to resist unhealthy and unwise choices [25].

Silvia Maier used brain images and showed that stressed individuals had more activity in the part of the brain involved in reward circuits and less activity in parts of the brain involving self-control [26]. The prefrontal cortex (System 2), which is needed for planning and decision making, is impaired when stress becomes high [27]. In such cases, people do not balance risks and are more impulsive in their decisions (defaulting to their System 1), making it even harder to look at the long-term goals needed to get out of poverty. It is a vicious cycle, not just because they are poor but also because their decisions work against them.

In another set of experiments, farmers were given tests before and after a harvest. Farmers are usually stressed before a harvest and are relieved afterward. From the tests, the farmers exhibited a clear improvement in cognitive abilities after the harvest.

And there's even more working against the poor. The rich will always get richer at the poor's expense. That is, money will travel from the large pool of poor people to the smaller pool of rich people. Here's the reason. The rich have advantages in acquiring wealth because they are rich. For instance, they pay lower interest rates on loans, have access to better financial advice, are better connected with other rich folks, can greatly influence laws and legislation to their advantage, and so on. But the poor have less time to shop for better prices, are under constant stress to make ends meet and therefore suffer from willpower depletion [28]. This is not a judgment of people's intentions. It is simply a proven reality.

Consider a game with everyone starting out with equal amounts of money and they each have an exact likelihood of winning. Each can bet as much as they like, and a coin flip determines the winner. On the first round, half will win and the other half will lose. The richer half now have more options and can leverage their increased wealth to give them a slight advantage—not much, but small advantages can easily propagate to large ones. Remember, the house has only a slight advantage in the roulette wheel game, but they almost always beat the pants off the players.

Sure, there will always be people who move between the two. Poor people who hit on an exceptional and marketable idea or just from luck or hard work can move upward. That's how some of our nation's immigrants do it. But in general, wealth travels upward from the bottom rung to the top.

All of which illustrates the challenges poorer nations will have in assisting global warming actions. Their populations are more concerned with surviving day to day than what will happen by 2050. It's not that they don't care. They do. But they care more about where their next meal is coming from. We would be no different.

Aliens and Dogs

SETI, the Search for Extraterrestrial Intelligence, is searching for signals from space that unambiguously come from intelligent aliens. Just confirming intelligent aliens are out there would be the scoop of the century. But a far more satisfying and useful accomplishment would be to communicate with them—chat them up if you will.

There are, of course, a few sticking points with this. The first is the time it will take to send and receive messages because of the vast distance between stars. The nearest potentially habitable planet orbits a star that is 100 light-years away [iii], so it will take 100 years to send them a message and another 100 years to receive an answer. No speed dating here.

[iii] A lightyear is the distance light travels in one year. Light moves at 186,000 miles per second so one lightyear is the same as 6 trillion miles. As a comparison, the Earth is 93 million miles from the sun which is a paltry 0.000016 lightyears.

Second, we would need to decipher their language or they ours.

But far more of an issue, I believe, would be understanding their Metabeliefs. And our newfound alien friends might have Metabeliefs unimaginably different from ours. Who knows what cultural, environmental, biological, and other differences they have that would drive their beliefs in directions incomprehensibly different from ours? To be sure, the laws of physics and chemistry, common to both of us, will limit the differences, but there could still be significant ones, especially since their technological level will likely be much different from ours. Trying to understand what they are saying would be impossible without knowing what they believe.

As an example, consider these two sentences (taken from the title of Lynn Truss's book [29]) in which only a single comma is added to the second.

Eats shoots and leaves.
Eats, shoots and leaves.

The first sentence seems to say that the subject (maybe a koala bear) eats plant shoots and leaves, its favorite foods. Whereas the second sentence seems to say that the subject (a hunter perhaps) eats and then shoots something (a gun perhaps), and then leaves the scene.

Look how easily the sentence's meaning changes—and this is from a single comma in a language and culture we are thoroughly familiar with. Have you ever tried sending a humorous email only to have it backfire because the nuances needed from body language and intonations were absent?

Now imagine the difficulties if we are dealing with a different everything, not just commas; different language, different culture, different history, and different biology. I think understanding them would be difficult and maybe impossible.

Which brings me to my favorite species, dogs. Ever wonder what goes through those little doggy brains? Quite a bit is my guess. Even dogs (or cats or chipmunks for that matter) must have Metabeliefs to survive day to day. Probably less sophisticated than ours but necessary nonetheless. Spot needs to know who his family members vs. who the strangers are that he needs to be wary of, if he is allowed on the couch, how to ask to go for a walk, and where his food and water bowls are. Further, dogs are more connected to humans than any other animal. They can sense our emotions and react with the appropriate behavior. It took us over 20,000 years to mold their Metabeliefs to our will, but now that we have, we are chummy BFFs.

How it all Hangs Together

The combination of Systems 1 and 2, our Metabeliefs, and confirmational biases work in concert to keep us alive and well. System 1 and System 2 are how we act on the world's stage—sometimes impulsively, sometimes thoughtfully. What guides them? Our Metabeliefs, which are everything we know and believe about the world and its inhabitants. System 1 automatically acts in accordance with those beliefs, never challenging them.

But the world doesn't always agree with our imperfect Metabeliefs. And when that happens, tension and conflict are created within our minds, which quickly spill over into our emotions. We get confused and upset and naturally fight to regain our balance.

So System 1 has the job of ridding us of these conflicts, thus righting the ship. It's similar to the way our immune system kills invading viruses. Left unchecked, the viruses will quickly overwhelm us, never a good thing for our longevity.

System 1 is the same and protects us from harm. One of its tools is confirmational bias, just as our immune system marshals its T cells. This command-and-control hierarchy is necessary for our day-to-day, even our second-to-second, survival.

But sometimes, just as an overcharged immune response kills off healthy cells, System 1 ignores or even kills off the truth, also not a good thing for our longevity. The cure? In both cases, turn down the immune system and turn down System 1. Engage System 2 and let critical thinking replace our instinctive, impulsive, and emotional decisions.

Connecting the Dots

1. Genes that don't help us survive and reproduce will be culled from humanity faster than Joey Chestnut can eat hotdogs. Okay, maybe not that fast, but on an evolutionary scale, fast enough.

2. We evolved two systems to deal with the world:

   • System 1: our autopilot that assesses threats and opportunities quickly and easily.

   • System 2: the smart kid side of us for careful and thoughtful contemplation when such is called for; never when it is not.

3. We mostly use System 1, but in our modern world, it can easily fool us,

especially because it is always looking for shortcuts.

4. Hucksters rely on System 1 to do our thinking. System 2 is the only tool we have to banish them.

5. But using System 2, which requires prolonged and intensive thought, is difficult, so we default to System 1 whenever we can, which is most of the time.

6. Just like a computer, Metabeliefs are our operating system that boots up whenever we start the day. It has everything we know and understand about the world around us and our place in it.

7. But our Metabeliefs can deny global warming is real if it contradicts our deeply held beliefs.

8. We all belong to groups, likely many of them. Groups help us through life and confer advantages of the many that we could not achieve on our own.

9. But groups also trap us into groupthink in which we believe everything the group believes without the desire or need to critically assess it.

10. We use two tools, among others, to protect our Metabeliefs and group norms from outside assaults.

  • Cognitive dissonance is the inability to simultaneously hold two opposing ideas. This leads to tension that our System 1 attempts to erase.

  • Confirmational bias is our dutiful gatekeeper and allows into our conscious mind only those facts that mesh with our long-held Metabeliefs. It ruthlessly eliminates facts that clash with those beliefs and, in so doing, releases the tension caused by cognitive dissonance.

11. But we are not helpless slaves to our evolved weaknesses. Forewarned is forearmed, so now we should be able to resist these debilitating tendencies. Chapter 3 provides additional tools to help.

Afterthought: Kick the Earth in the Butt

"You can kick the [earth] in the butt really really hard and it will come back."

—David Droegemeier, White House director of science and technology on his global warming views [30]

Droegemeier is right. The earth will come back. The problem is *we* will not. The earth will get along swimmingly once we are gone. The good news, at least, is that we will not have to put up with such absurd statements.

References

[1]   Kahneman D. Thinking Fast and Slow. New York: Farrar, Straus and Giroux 2011.

[2]   Willingham DT. When You Can Trust the Experts. 2012.

[3]   Kahneman D. Thinking Fast and Slow. New York: Farrar, Straus and Giroux 2011; p. 101.

[4]   Willingham DT. When You Can Trust the Experts. San Francisco: Jossey-Bass 2012; p. 45.

[5]   Sharkey N. Autonomous Warfare. Sci Am 2020.

[6]   Formi PM. The Thinking Life. New York: St. Martin's Press 2011; p. 11.

[7]   DeStano D. The Only Way to Keep Your ResolutionsNew York Times. 2017.

[8]   Fields R. The Roots of Human Aggression. Sci Am 2019.
        [PMID: 34276077]

[9]   Gingras C. Is there a rough estimation of how much data the human brain receives per second?Quora.

[10]   Chew L S. Myth: Eyewitness evidence is the best kind of evidence. Association for Psychological Science 2018.

[11]   Lewis M. Risk vs. Risk: California Farmers Worry More about Climate Policy Than Climate Change. GlobalWarmingorg 2013.

[12]   Federico CM. Us Against Them. Science 2018; 362(6413): 26.

[13]   Quoted in Michael Shermer. For the Love of Science. Sci Am 2018.

[14]   Butts CT. Why I know but don't believe. Science 2016; 354(6310): 286-7.
        [http://dx.doi.org/10.1126/science.aaj1817] [PMID: 27846517]

[15]   Jonestown. Wikipedia Jonestown

[16]   Hogg M. Radical Change. Sci Am 2019.

[17]   O'Connor C, Weatherall JO. Why We Trust Lies. Sci Am 2019.

[18]   Kurtzleben D. Surprised about Donald Trump's Popularity? You Shouldn't Be. NPR 2015.

[19]   Willingham DT. When Can You Trust the Experts?. San Francisco: Jossey-Bass 2012; p. 82.

[20]   Herrera T. So You've Made a Huge Mistake? What Now? New York Times 2019.

[21]   Gorman S, Gorman J. Denying to the Grave: Why We Ignore the Facts That Will Save Us. Oxford University Press 2016; p. 149.
        [http://dx.doi.org/10.1093/oso/9780199396603.001.0001]

[22]   Otto S. A Plan to Defend against the War on Science. Sci Am 2016.

[23]   Lindberg E. Six Myths about Climate Change That Liberals Rarely Question un-Denial 2016.

[24]   Danziger S, Levav J, Avnaim-Pesso L. Extraneous Factors in Judicial Decisions Proceedings of the National Academy of Sciences.
        [http://dx.doi.org/10.1073/pnas.1018033108]

[25]   Vohs KD. Psychology. The poor's poor mental power. Science 2013; 341(6149): 969-70.
        [http://dx.doi.org/10.1126/science.1244172] [PMID: 23990551]

[26]   Maier SU, Makwana AB, Hare TA. Acute Stress Impairs Self-Control in Goal-Directed Choice by Altering Multiple Functional Connections within the Brain's Decision Circuits. Neuron 2015; 87(3): 621-31.
        [http://dx.doi.org/10.1016/j.neuron.2015.07.005] [PMID: 26247866]

[27]   Sapolsky RM. The Health-Wealth Gap. Sci Am 2018; 319(5): 62-7.
        [http://dx.doi.org/10.1038/scientificamerican1118-62] [PMID: 30328837]

[28]   Boghosian B. The Inescapable Casino. Sci Am 2019.

[29]   Truss L. Eats, Shoots & Leaves, the Zero Tolerance Approach to Punctuation. New York: Gotham Books 2003.

[30]   Nominee Ducks Climate Query. Science 2018; 261(6405): 31.

CHAPTER 2

Only through their Beliefs and Emotions can we Reach their Minds

What a man had rather were true, he more readily believes.

—*Sir Francis Bacon, 1868*

The NY Yankees baseball team has won a quarter of all possible championships, an astounding feat unmatched by any team in any major sport. So in light of this indisputable fact, tell the next Bostonian you meet to drop the Sox and root for the Yanks. You can guess the outcome, and it's not going to be pretty.

In trying to convince someone, facts will not work if those facts challenge a person's Metabeliefs or their sense of who they are especially as it relates to their friends, relatives, and others they admire. As Michael Shermer writes, "People with more knowledge only accept [facts] when it doesn't conflict with their preexisting beliefs and values. Otherwise they use that knowledge to more strongly justify their own positions [1]".

Facts can be ignored, changed, or bent to anyone's purpose. For instance, look at people who believe the earth is flat. They go to great lengths to overcome the obvious facts arrayed against them, and do so with extraordinary oblivion to the real world: "They believe the Earth is a disc with the Arctic Circle in the center and Antarctica, a 150-foot-tall wall of ice, around the rim. NASA employees, they say, guard this ice wall to prevent people from climbing over and falling off the disc [2]". You can't make this shit up.

The way a person decides on major issues is the same way they decide on their sports team. Their thoughts are shaped by their Metabeliefs, their friends and relatives, where they live, and so on. Facts may come into play but only as bit actors in the background of a much larger unfolding drama.

This leads to a problem with consequences that will shape history and our well-being, if not our survival. We must convince people to take action, but doing it only with facts is like throwing pebbles at a charging rhino. It doesn't work. Actually, it leaves us worse off, since facts, true or not, will only harden a person's convictions, leaving them even less susceptible to change, especially (as Katharine Hayhoe notes) if they see it as an attack on their identity and way of life. We are just annoying the rhino [3].

So to convince folks about the need for global warming action, we must tailor our message to fit in with their beliefs, how they view the world, how they think of themselves, what is important to them, and what their hot buttons are.

This isn't going to work all the time and maybe not even most of the time, but hopefully often enough to be successful, the same way a vaccine provides herd immunity: if enough people take action, the rest will be saved.

CHAPTER CONTENTS

Friend or Foe?

In the following, put the person in the correct column.

Why is this exercise important? Why must we know the positions people take on global warming? Imagine Wile E. Coyote mistakenly walking on stage to an audience of Road Runners all holding boxes of Acme dynamite. It really, really pays to know our audience.

What are their Global Warming Beliefs?

Person	Will Likely Support Global Warming Action	Will Likely Not Support Global Warming Action	Cannot Be Determined
Black Female			
White Protestant			
Reads *Wall Street Journal*			
White, Attended College, Not Religious			
White and Gay			
White and Jewish			
Has a College Degree			
Asian and Married			
Asian Protestant			
Black Male under 31			
Older White Male			
Unmarried Female			
White Male Did Not Attend College and Not Religious			
Watches Fox News			
Lives in New York			
Lives in West Virginia			
Wealthy Religious Hispanic			
Pope Francis			

The same with global warming. It's important to know if we are dealing with Wile E. Coyote or Road Runner and what their global warming beliefs are. If they are certain that global warming is happening, then the dialog should center on the best way to mitigate it. If they are on the fence, then convincing them takes the center stage. Finally, if they are not going to budge, then we either abandon them or play to the crowd that is listening. Fortunately, it is possible to get insight into a person's global warming beliefs through three simple traits:

• Their political and religious beliefs

• Where do they get their news from?

• Where do they live?

Political and Religious Beliefs

Political beliefs strongly affect a person's stand on global warming, perhaps more so than any other trait [4]. Conservative Republicans are much more likely to disbelieve global warming is happening, while liberal Democrats take the opposite stand. Even among the conservatives who accept global warming, they still resist the notion that we are responsible, believing it is caused by natural events. And each position has strengthened over the past 10 years [5], [6].

Surprisingly, it is not difficult to determine a person's political stance and therefore their global warming beliefs. Sahil Chinoy reports on research in which 140,000 people were surveyed about their political leanings [7]. Two important characteristics are a person's race and religion. If you are black, Hispanic, or Asian, you are more likely to be a Democrat and support global warming actions, whereas if you are white and religious, you are more likely to be a Republican and deny global warming. The following table shows how different traits will portend a definite lean.

Global Warming Beliefs

	Democratic Lean (Favors Global Warming Action)	Republican Lean (Opposes Global Warming Action)
Religious		X
Gay	X	
College Educated	X	
Older		X
Lives in the South		X
Lives on Either Coast	X	
Black, Asian, Hispanic	X	
White and Religious		X
White Not Religious	X	
Millennial Women	X	

Where They Get Their News

If you get your news from the following, you are likely to be conservative and have less trust in global warming and will be influenced by the anti-global warming reporting:

• *Breitbart News*

• *Daily Mail*

• *Daily Signal*

• *Drudge Report*

• *Fox News*

• *National Enquirer*

• *The Blaze*

• *The Federalist*

• *The Glen Beck Program*

• *The Rush Limbaugh Show*

• *The Sean Hannity Show*

• *Washington Examiner*

• *Weekly Standard*

Whereas if you read, watch, or listen to the following, you are likely to be liberal and will favor global warming sentiments and are influenced by their pro-global warming reporting:

• *ABC News*

• *BBC*

• *Bloomberg*

• *CNN*

• *Google News*

• *Mother Jones*

- *MSNBC*

- *NBC News*

- *NPR*

- *New York Times*

- *PBS*

- *The Daily Show*

- *The Guardian*

- *The Huffington Post*

- *The New Yorker*

- *US Today*

- *Washington Post*

- *Yahoo News*

The *Wall Street Journal* is one of the only news sites that has a loyal following of both liberals and conservatives, so it is on the global warming fence.

Where They Live

Yale conducted a comprehensive study that looked at people's global warming opinions starting at the national level and drilling down to state, congressional districts, metro areas, and even down to the county level [8]. This is as fine a detail as you will find anywhere.

In general, the West and East coasts tend to be pro-global warming action while the Midwest, Northwest, and Southwest tend to lag behind these sentiments.

And contrary to what some say, size does matter. For global warming anyway. A population density of around one thousand people per square mile seems to separate global warming advocates (higher density) from those opposed (lower density) based on their political affiliations [9i]. Living in sleepy Mayberry, no matter who the sheriff is, is much different that in bustling Atlanta. Urban areas are likely in favor of global warming action whereas rural areas tend not to be.

---

[i] © 2019 David Troy (@davetroy). Source: 2010 US 2015 Census Community Survey. Election Data politico.com, excludes third parties.

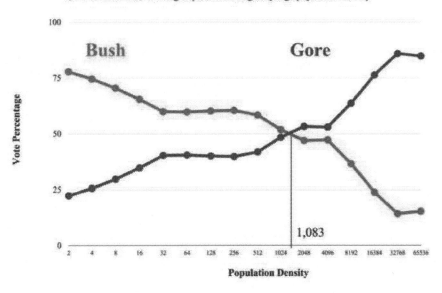

(a)

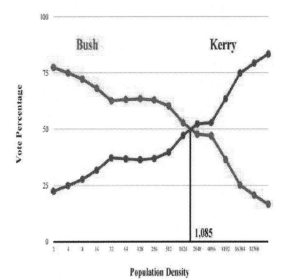

(b)

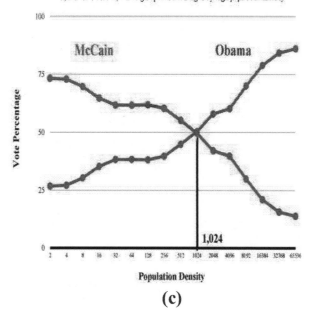

(c)

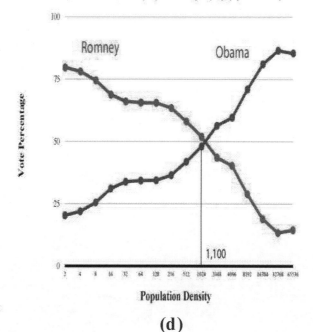

(d)

In this fashion, a person's global warming views can usually and quickly be determined, though it is by no means flawless. Of course, the most direct approach is simply to ask, "Do you think action on global warming is needed?" This is not always possible, as we may be addressing a crowd, but getting a sense of where they stand is important in crafting our approach.

We don't ever want to walk onto a stage with Road Runners staring us down.

To Fear or not to Fear?

Do we shock people out of their complacency with tales of global warming doom or do we take a more paternal tack and lead them to what needs to be done? Research supports taking a kinder, gentler approach.

A study looking into convincing parents to have their children vaccinated by confronting them with scary pictures and a tragic story not only didn't work out but also backfired and hardened the convictions of those parents already against vaccinations [10].

Further, anxiety and fear allow conspiracy theories to flourish as they provide comfort in making all problems, including global warming, disappear in a fog of lies and half-truths, with the problem misdirected against another target, say scientists, which then stops global warming action in its tracks [11].

In ancient times, the bearer of bad news was usually executed—and for good reason. If your day just got ruined by someone pointing out the angry peasants amassing at your castle gate with pitchforks and torches, your instinct is to strike back at the source of the pain. Nowadays, society frowns upon killing the messenger, but the psychological effect remains, and we will strike back by not believing the message and withdrawing further into our cocoon of denial. It is as natural as spitting out a bitter-tasting food.

But offer a stranger a chocolate bar and you will be showered with smiles and goodwill, open to anything you wish to tell them.

Our path is obvious. We must lower people's fear and approach them with realistic and doable actions, especially if they also can reap some benefit. People must feel they have some control over the problem and that actions are within their abilities and best interests.

As Gelspan pointed out, Martin Luther King Jr. gave the "I have a dream" speech, not the "I have a nightmare" speech. So must we.

The Truth Shall Set You Free—John 8:31-32

A well-known global warming advocate stated that nuclear energy was much safer than fossil fuel-fired electric generation and further stated that no one has ever died from nuclear plants whereas many have died as a result of fossil-fired plants.[ii] This is blatantly wrong on many levels. First, it can be easily debunked by just googling "deaths from nuclear power," which shows fifty deaths from Chernobyl and about 1,450 deaths from Fukushima [12].

Second, there are sixty nuclear plants in the U.S. compared to over two thousand fossil-fired plants, so comparing the deaths between them is like comparing the road fatalities in a sleepy village to that of a large metropolitan city. Yes, there are many more driving deaths in the large city, but that's because there are many more drivers than in the village, not because the village is any safer.

Certainly, there are studies that postulate that nuclear energy is safer than fossil fuel–fired energy, but the assumptions used make their accuracy debatable. As an important example, the number of deaths that will occur due to storing highly radioactive waste for one million years was not addressed, nor really could it be, except maybe in a philosophically broad-brush way. The correct statement should be "We really don't know how many deaths will occur due to the widespread use of nuclear energy, but it should be one of the tools we use to combat global warming, and we must constantly improve its safety as our understanding and the technology advances." And not "nuclear is safer than fossil-fired plants." Let's be upfront.

Here's another road better less traveled. An advocate for global warming action suggested we withhold unpleasant facts to further our cause: "In politics, if you show uncertainty . . . you may be at a disadvantage, and you may need to do some soul searching about how much uncertainty you hold back [13]". I realize our approach needs to be tailored to each situation, but I am opposed to coloring the truth by intentional omissions, which will always come back and bite us. Now, I don't suggest we raise unimportant issues just to show we are fair, but we should bring up items that are germane to our audience's interests, and if uncertainty (or any other issue) plays into this, then we need to deal with it head-on. If our audience misunderstands uncertainty, then give them a quick tutorial of how uncertainty permeates our lives everywhere, is uncompromising, and should not be feared but accepted (see chapter 9).

Intentionally withholding relevant facts is deceptive and is worse than an outright lie, since half-truths are harder to detect.

[ii] Fossil fuels refer to natural gas, coal, oil, gasoline, propane, and some other minor ones.

Some folks promise that aggressive actions will halt and reverse global warming and we will then be back to the pristine world we knew as children. All will be forgiven and our happy carefree days will return. Whether actually stated or implied, these are misleading and false. We are never going back, as irreversible global warming damage has already occurred, and even if we miraculously stop all greenhouse gas emissions, the sheer inertia of what we have started will continue to warm the world for a good while. Don't go there and don't allow anyone else to do so.

A friend of mine accused Representative Ilhan Omar (D-MN) of supporting al-Qaeda. This was clearly a false statement [14], yet my friend did not fact-check it, probably knew it was false, but chose to gleefully pass it on anyway. I suspect this goes beyond confirmational bias and lands in the realm of willfully disseminating falsehoods because it supports a position and likely gives some gratification.

I always thought (maybe hoped) that truth and facts were the final arbiters of any argument, but now these two seemingly old-fashioned ideas are just tools to be contorted into any desired position and wielded as a bludgeon against all who hold an opposing view. The news abounds with such travesties. When we let these lies perpetuate, we are doing far more damage to our country than just denying global warming. We are allowing counterfeit facts to upstage and replace the truth because it is more appealing or glitzier, or it provides some pleasure within our ancient brain stem. But to do so, and not to call out others who do so, is playing a dangerous game in which we can be manipulated by any skilled orator without the protection of the truth to guide us. It is likely one of the greatest threats to our culture.

And this leads to a much wider problem, as once we become insensitive to lies, it not only becomes easier, but it also moves us toward believing lies are necessary to keep up with our opponents' onslaught. History will judge us all; it is as inevitable as hot summer days. Let's not have history look upon our generation as those who destroyed the truth, no matter how noble the cause is.

Truth is precious, so please don't make it ephemeral as well. It serves us much better when people know we can be trusted.

Always Take the High Road

A gym member continually annoyed me with his loud cell phone conversations. Admittedly, I annoy easily, one of my minor demons, but he was over the top with his attention grabbing. One day he mounted the elliptical next to mine and I worried, what is this peacock going to do now? Instead, he leaned over to me and said, "Nice workout." My resistance collapsed like a house of greased cards and I haven't thought poorly of him since. What did he do? He took the high road and disarmed me, not with confrontation but with friendship and interest. And he reinforced an important lesson for me.

Treat your opponents and allies the same. Engage them with respect to win them over, which makes them receptive to what we say. Try to become one of them, at least in understanding what drives their opinions. Find common ground, whenever you can to make that personal connection. And remember that people will be more receptive to new ideas when they feel good about themselves and are not under a siege of threats. Offer them a psychological candy bar.

I was in a global warming discussion with a religious white male conservative. He had no interest in learning but was just trying to win debating points and used the old tried-and-true "global warming is just a theory" approach, wrongly implying the word "theory" means it is not proven but just someone's half-baked idea. My demon raised its ugly head, and I snapped back, "Gravity is just a theory, too, so why don't you jump off a ten-story building and tell me how that works out for you." Whatever slim chance I had of convincing him went down in flames.

I have seen this many times: global warming advocates waving the flag of righteousness with truth on their side riding valiantly into the global warming battle. In fact, it might as well be the Charge of the Light Brigade, as they convince no one and indeed make matters worse with their obnoxious "I'm better than thou, you unwashed heathen" attitude.

Instead, we need to listen to our opponents. They have a side to tell, and only by listening can we get the lay of the land on them. My brother Paul, a salesman extraordinaire, explained that the sales job doesn't begin until you get the first no. Find the no's our opponent is expressing, not to counter them so much but to understand their Metabeliefs, which can provide a road map to their hearts.

Most of all, we need to be honest with ourselves. Why are we in this game? Are we just trying to win the argument, or are we really concerned and trying to alert everyone that the global warming Kraken has been released upon us? I was at a global warming seminar and talked to a person advocating a carbon tax to reduce emissions. I was interested in learning more, so I asked him what the problems are with implementing this. He responded by attacking a position he thought I held (which I didn't) and aggressively tried to get me into a five-hundred-dollar bet that he was right. Get enough of those types on the streets and we might as well climb into a deep hole and wait for the End Times. He was only interested in winning the argument and likely goes about undoing the work of others who push for global warming actions.

We must resist our instincts to lash out at those who refuse to believe or who fire head-exploding illogical arguments at us. Stay calm and stay the course even when it goes against our nature. Tame our ego.

I am sensing my readers' unease with this, so go ahead and unburden yourself.

"You are asking me to suppress my ego, but isn't my ego part of my rigid Metabeliefs that you keep droning on about? How can I change, since, as you have amply demonstrated, we should go with, not try to thwart, one's Metabeliefs?"

Point well taken. We can't change our stripes, and I am not suggesting we submerge our egos, but rather to redirect them away from who we are and onto what we can accomplish. Our accomplishments become an indelible force on our Metabeliefs that define us. We will not be perfect at it; no one can ever be. Remember my own demon that turns my otherwise civil conversations into hash whenever I let my guard down. But in the name of our survival, let us adapt to the situation as best we can.

Compromising is A Powerful Tool—Use It

We are trying to save the world from the catastrophic effects of global warming, so the only thing we need to do is convince people to take action on mitigating global warming effects. That's the ball, and the only ball, we need to keep our eye on.

I know someone who finally agreed, after many years of denial, that global warming was happening, but it wasn't caused by us but by something the earth was traveling through. I'll accept that because he moved to an important position. And don't think that was easy for him. It took courage and intellect to modify his Metabeliefs. Showing this much movement with someone who had deep conservative views is amazing and is a testament to his ability to adapt.

Besides, one step at a time. They will come around eventually as their Metabeliefs continue to evolve. Asking for too much too soon will lead them to mental fatigue and resistance.

The all-or-nothing, take-no-prisoners attitude that some global warming enthusiasts have serve no one's purpose and jeopardize humanity's well-being no less than global warming deniers do.

Be prepared to compromise and stay focused on the goal. We don't want to bend our opponents' will, but rather get them to take action. We should give in on minor points so as to achieve the overall aim to save humanity. Compromise is always good as long as we attain the key issue—taking action on global warming. And it gives our opponents a way of saving face.

Time is Against us

At some point we may need to play the long game and just wait for the current generation to fade into the sunset and work on the younger kids who might be more open-minded to global warming issues and who will also be far more affected than their parents.

In fact, among young adults (18 to 29 years old) in 2019, over 45 percent felt that the federal government should step in to mitigate global warming effects [15], compared to only 31 percent in 2015. That's a 30 percent increase in just four years.

So it may take a generation or two for global warming action to be accepted across all population segments.

Unfortunately, time is not on our side. The Intergovernmental Panel on Climate Change (IPCC) has stated that we have to slash our greenhouse gas emissions by 45 percent by 2030 and down to zero by 2050 to keep our global temperature rise to below 1.5°C (2.7°F) [16]. Going above this will cause extensive damage, and the higher above it, the more catastrophic it becomes. This is a Herculean task even if we start today, but if we wait, it will become a daunting and perhaps unpayable mortgage we are crushing our children with.

Action is needed now. We can't wait to figure out all the details or be comfortable in our own skins that this is the right approach. We must start, and start quickly.

"As an Aside, speaking of time, ever wonder how long the present is? Is it a microsecond, a picosecond, or something else? If we are moving from one moment to the next, from the present to the future, where are we at this instant? One moment we are in the present, which then skips into the past without our noticing. Oops, there goes another one".

"But what about right now, this very instant of time as you read this last word? It would seem our lives—all of reality, in fact—exist on a seemingly paper-thin boundary between the past and the future. A blink of an eye between the nearly infinite time that came before and will come after.".

"We are aware only of the present, that tiniest instant of time we spend our whole lives in. We have no awareness of the past, just a vague memory of what it was. We have neither an awareness nor a memory of the future, just a guess of what is to come and maybe some certainty that it is there, waiting for us but never materializing. We are just inches from it, moving toward it, yet it remains always maddingly out of our reach. It's like a dream where we are forever running but never reaching our goal".

"Anyway, time is doing its thing, rushing off to the next moment, stopping for no one. So let's get back to the matter at hand and move on to the next section".

Will It Take A Cataclysm?

It may take a global warming cataclysm to knock people out of their complacency, since we evolved to respond to a world where the dangers are immediate and visible.

Global warming is abstract and distant, not part of today's more pressing problems. And it is at the bottom of an almost bottomless pit of worries with more near-term and visible problems such as paying the rent, caring for our elderly parents, and terrorist attacks—all of which supersede the vagaries of global warming.

This gets back to people's notion of time. The fact that the Florida Keys will be underwater by the end of the century will not activate many brain neurons, as people care more about the present and not so much about the future. People react to immediate threats quickly, short-term threats somewhat, and long-term threats maybe not at all, which is perfectly sensible. There is no sense worrying about long-term threats if we don't first survive the immediate ones.

Extreme weather gives some credence to the cataclysm theory. Twenty-one percent of people taking a survey said they changed their minds about global warming because of the extreme weather they are experiencing: "People's perception of risk is heightened after a major hurricane or wildfire [17]".

Another example is when the comet Shoemaker Levy 9 dramatically smashed into Jupiter, which was photographed and widely reported in the press. It got everyone's attention, and while there always has been concern about chunks of rocks hitting earth and making us go the way of the dinosaurs, this starkly visual cataclysm precipitated urgent demands for swift action to protect our world.

This brings us to the unfortunate need for an immediate wake-up call that is dramatic and may be dire. What we will need is a cataclysmic global warming event that is obvious and close to home. I'm sorry to say there are plenty to choose from and more coming our way. Some examples are monster storms rarely if ever experienced before, droughts that damage crops and people's income, flooding, storm surges, increasing forest fires (look at what California is going through), and extreme weather, to mention a few. And they are surely coming for us. That's also as inevitable as hot summer days.

So let's always be prepared for some global warming calamity to hit the front pages. Then we can go on the offensive and stalk our targets.

Don't Underestimate the Enemy

Even with all the convincing we do, the deniers and their ilk may take surprising actions to accommodate their Metabeliefs. An off-the-chart example was the Oregon State Republicans who left the state after the state's Democrats introduced a bill to cut carbon emissions. The Republicans knew they would lose the vote, so they skedaddled to be sure a quorum was not possible and therefore no action on the bill could happen. But it gets worse. They threatened to shoot anyone pursuing them, or as Senator Brian Boquist told the police, "Send bachelors and come heavily armed [18]".

A good place to get a feel as to how deniers will react is to take a peek at how the anti-vaxxers' arguments evolved. These are the parents opposed to vaccinating their children. First, they thought vaccinations caused autism from a seriously flawed paper by Andrew Wakefield, who preyed on the historical hesitation to vaccinations in general. It was shown that he falsified his research results because he was investing in an alternate vaccine [19]. Once outed, he lost his medical license and was disgraced, but his lies worked their way into the anti-vaxxers Metabeliefs and they could not let it go.

So the anti-vaccine faction moved the narrative to a different battleground, one that involves self-determination and freedom from unwarranted government intrusions into their personal lives, as well as religious freedom. And you can bet global warming deniers are going to use the same playbook, especially since suspicion of government interference has always been part of their lexicon.

Beliefs seem to take on a life of their own, replete with strong self-survival instincts. They refuse to die no matter the reality and will go to extremes to keep perpetuating their offspring. Conspiracy theories are good examples. Once they get a foothold in the wacky world of social media, there is no stopping them. Area 51, with its cast of aliens, spaceships, mad scientists, and government cover-ups, is one of many examples that are never going away. It is nearly immortal and certainly immune to any logical attacks and will live on long after we have departed.

It appears illogical, I know, but there's nothing to be done except acknowledging the phenomenon and adapting to it.

Convincing Stereotypes

Generally Speaking, It's Their Metabeliefs

In the following sections, seven different stereotypes have been selected as examples of what approaches might work. And as you may have guessed, a person's Metabeliefs stand above all others. People will reject evidence of global warming, however true it might be, if it attacks beliefs they and their group hold. If you have followed my jabberings till now then I am sure you will embrace this point. Or perhaps you are using the book for target practice. Either way is fine as long as you paid full price. But as usual, I seem to be drifting so let me get back to my monolog.

And remember, a person's Metabeliefs are not all or nothing but consist of a rich variety of issues that constantly adjust to a single stable system, the way water finds its own level. So under the surface there will always be a tug of war among disparate beliefs that the mind constantly rationalizes into the whole. But push a little harder on one of these and the entire belief system might budge a bit in the direction we want. The amount of push depends on how open a person is to new ideas. Those set in their ways may never move, and those too easily convinced will be constantly oscillating between this belief and that belief depending on who talked to them last. Both types are untenable targets and should be avoided. What we want is the Goldilocks Metabeliefs that are malleable but not too much. Those open to new ideas, at least on this topic, are ideal.

An approach that aligns with a person's Metabeliefs is when one of their own provides the convincing. Having someone from the military discuss this with conservatives, or having a religious leader discuss it with their congregations are two examples. Or having influencers on social media support global warming positions. If we can get to them, that would be a bull's eye with considerable effect.

Developing a connection with our target is essential. Something shared, such as both being parents and worried about their children's future, rooting for the same sports team, which, according to Jay Van Bavel [20], will supersede any political differences, since the team always comes first.

People prefer stories they can relate to, not abstract ideas and complicated data, which will lose them as fast as your footing on an icy sidewalk. Wendy Zukerman points out that arguments that are counterintuitive, that go against what someone would view as common sense, will not work no matter how well structured or true [21]. This is a person's System 1 in action doing its job of protecting its host with minimal effort and energy. To satisfy these System 1 safeguards, make an emotional story that resonates with their Metabeliefs.

For example, I was lecturing on confirmational bias and I illustrated its use with a heart-wrenching story about a mom who lost her child to a quack doctor but who nonetheless defended the doctor so as not to have to share the blame of choosing him in the first place. You could have heard a pin drop in the room.

Keep it simple and uncomplicated so their System 1 can absorb it effortlessly. This will resonate to people by appealing to their instincts. Make it fun and sexy. Use stories that connect with the audience. Avoid facts except to support an emotional appeal. Anecdotes, personal stories, even humor works if the timing is good.

"It's the economy" wins presidential elections. "It's their Metabeliefs" wins arguments.

Religious

Fifteen percent of Americans believe God controls climate and people can't influence it [22]. There are 200 million adults aged 18 to 64 in the U.S. so that is 30 million people. Imagine that. Thirty million people believe we can sleep snug as a bug in a rug and a divine power will wash away global warming without any action on our part.

Religion is a significant force in people's lives, maybe the most significant, and greatly shapes their Metabeliefs, which influence their stand on many other issues besides religious ones. For instance, being opposed to evolution might provide arguments that the climate cannot similarly change without divine approval, leading to the snug-as-a-bug-in-a-rug feeling. Inaction, unfortunately, is the outcome.

But in fact most religions teach otherwise. The pope embraces the scientific view of global warming and acknowledges that global warming is occurring and we are responsible for it:

"A very solid scientific consensus indicates that we are presently witnessing a disturbing warming of the climate system."

"Humanity is called to . . . combat this warming."

"A number of scientific studies indicate . . . global warming is due to greenhouse gases . . . mainly as a result of human activity"[23].

He also quotes Patriarch Bartholomew:

"For human beings to degrade the integrity of the earth by causing changes in its climate . . . these are sins. For to commit a crime against the natural works is a sin against ourselves and is a sin against God."

Religious support is not confined to the Catholic Church. The Church of England is also backing global warming actions by divesting their financial holdings from companies not doing enough to combat global warming [24].

And many other religions are including global warming action statements [25]. The Yale Forum on Religion and Ecology lists the following religions, among others, that support global warming actions:

• *Buddhism*

"Today it is our responsibility as Buddhists and as human beings to respond to an unfolding human-made climate emergency that threatens life. There is an uncontestable scientific consensus that our addiction to fossil fuels and the resulting release of massive amounts of carbon has already reached a tipping point."

• *Hinduism*

"Climate change creates pain, suffering, and violence. Unless we change how we use energy, how we use the land, how we grow our crops, how we treat other animals, and how we use natural resources, we will only further this pain, suffering, and violence".

• *Islam*

"No one is exempt from the vagaries of climate change and Muslims have to accept their share of the responsibility for bringing this onto ourselves".

• *Judaism*

"Major sectors of the Jewish Community are taking strong positions on combating climate change"

But it's complicated. There are numerous forces either acting in concert or opposing people's religious beliefs. Religions that value nature and its protection could be overridden by a desire for a higher living standard, by war, indifference, or fear of other undesirable ideas tagging along. So despite the above statements, religious folks may still deny global warming. For instance, white male Protestants tend to doubt global warming, and their disbelief goes deeper if they are also straight, older, or from the South. And only 28 percent of evangelicals believe global warming is caused by people, compared to 50 percent of all U.S. adults [26].

Why? As in all things, their Metabeliefs assert themselves like a body snatcher driving their actions and thoughts. As pointed out by Hayhoe, "Religious beliefs can be interpreted to support conservative views on climate change." Or as Congressman Tim Walberg said in 2017, "As a Christian, I believe that there is a creator in God who is much bigger than us. And I'm confident that, if there's a real problem, He can take care of it."

Plus, understanding global warming, like any complex topic, takes effort and focus. The religious might tune out any global warming facts, not because they can't or don't want to understand it but because they just don't put the time into it. Why study something you have already dismissed?

What's to be done? It depends

Religions fall into two broad environmental categories: those that embrace our oneness with nature and those that have us as both masters of nature and apart from it [27]. Eastern religions such as Buddhism emphasize the connection and inseparability of people and nature, whereas Western religions of Judeo-Christian origins teach that nature is for our use and consumption.

One approach, then, is to stress that the Bible and the Koran teach us to be stewards of the planet respecting His divine creation, and thus we have a responsibility to protect our planet the same way parents must protect their children. And demonstrating is always more powerful than talking at people. Take them on a walk through a park and show them God's creations that we are entrusted to protect.

And for those snug as-a-bug in-a-rug folks, God has always been content to let us do our own thing and face the consequences. Where was He during the 1918 Spanish flu, the bubonic plague, or when millions of civilians were slaughtered in World War II? And let's not forget His own disasters: Noah's ark, the Seven Plagues, and the original—throwing Adam and Eve out of Paradise. Where was He? He was always there, allowing free will to show our true colors. The same applies with global warming. We need to take responsibility and protect what God has given us.

A complementary approach is to stress that fighting climate change is also fighting social ills and creating a better life for our children.

It's always a good idea to work with the alpha dogs: the priests, rabbis, ministers, and imams that lead the congregations. They are the gatekeepers and very influential, as any leader of a group would be. We can move mountains with the religious leaders on our side because of the influence they have and the normal group dynamics that favor a respected articulate leader. Plus, leaders tend to be concerned more about their flocks and less about themselves, which makes them more open-minded to global problems. They are also more likely to be attuned to their religious leadership's direction on global warming, which would counterbalance opposing secular concerns.

Also, fighting global warming yields great PR for the religion and may even be a form of evangelizing to gain more converts. However, be careful here of crossing the reciprocity line, in which someone feels obligated to convert because of the services being offered by the religious group or because of the authority the religious group represents.

Evangelicals are a breed apart since they make up 30 percent of the U.S. population and are politically active and are thus a very desirable voting bloc to address [28]. They live mostly in conservative states, with about half of them in the South, a bit less than one-quarter in both the Midwest and West, and a smaller number in the Northeast. And consistent with where they live, they mostly vote conservatively.

Evangelicals may resist global warming because they fear it will lead to other liberal ideas, such as abortion and same-sex marriage [29], issues that are critically important to them. It becomes a slippery slope in which once you acknowledge global warming, you open the door to other liberal ideas that gain speed too fast to halt. Hence, an important and necessary approach for us is to uncouple global warming from the liberal agenda. Otherwise, we will be fighting a two-front war, that of global warming and that of other liberal values, which assures failure.

Global warming must stand on its own, independent of other ideologies, and this must be made clear. Politicians who support global warming action but also push a liberal agenda are going to have rough sledding with evangelicals if they cannot separate the two. Conversely, conservative politicians are favored by evangelicals, but they will not take action on global warming.

However, cracks are appearing in the evangelicals' resistance to global warming action. Take, for instance, the Evangelical Environmental Network, which has 4 million members. They are pushing for action on global warming for the children's sake. As they state, "Free our children from pollution by relying entirely on clean electricity from renewable resources like wind and solar by 2030."

Further, young evangelicals are also taking a positive stand on global warming, through their Young Evangelicals for Climate Action (YECA) organization: "We believe God is calling us to faithful action and witness in the midst of the current climate crisis [30]".

So there is hope, but it is an uphill climb, as it takes time for people to come around to new ideas, no matter how important and pressing they are.

Conservatives

Conservatives tend to be religious and favor authority, hierarchy, limited government interference, and a strong military, among other traits. Generally, they are suspicious of global warming and, more so, that we are responsible for it. Hence, they do not see global warming action as desirable.

Again, it is complicated. Conservatives may oppose global warming, not necessarily because of a disbelief in it but because of their fear of government overreach. To them, global warming is just another tool to be used by the federal government to control them. Hence, the entirety of the global warming debates becomes a proxy for their push-back on government interference in their lives.

Further, researchers have also shown that conservatives reject global warming, again not because they don't understand the facts but because they get "their cues from conservative elites, many of whom have close ties to the fossil fuel industry [31]".

Add to this their strong religious beliefs, also complicated, and you get a mosaic of influencing forces pushing conservatives away from global warming action.

What's to be done? First of all, respect conservative views and beliefs that are based on sound experience and tradition and have served them well over the years. Trying to convince conservatives to embrace the love and warmth of government bureaucrats is no better than trying to convince liberals to put down their lattes and go line dancing. It's like teaching a cat to fly by throwing it in the air. It doesn't work and it annoys the cat.

Work with them so that global warming action it is to their benefit. I offer a few tips.

Conservatives will be put off by carbon limits if they believe it will constrain industry. But if global warming solutions such as nuclear and geoengineering are mentioned, the context shifts to having industry lead the charge, and that will appeal to them [32]. Clean energy solutions are already benefiting the middle of the country, with Texas having 233,000 jobs in the clean energy industry, second only to California [33], [34].

Conservatives have a deep respect for the military. And the military is quite concerned with the effects global warming will have on its own operations and the U.S.'s security, as I discuss in chapter 10. Conservatives should take notice that the military accept global warming, and are also addressing its consequences. To support the military is to support their missions, which means supporting their efforts on global warming.

But tread carefully here if you are not currently or formally from the military yourself; otherwise your credibility will be suspect and your attempts may fail.

Conservatives are fiercely independent and often distrust the federal government, so an important approach is to put any action into our own hands and let market forces determine the most efficient way to handle it. Research by Cary Funk and Brian Kennedy has shown that conservatives would rather have the general public make global warming policy decisions rather than politicians, climate scientists, and certainly not other countries [35].

So in that context we should also be wary of foreign governments getting the upper hand in global warming action and should certainly not let them reap the financial rewards of it or allow them to dictate U.S. policy. Changing the narrative from the U.S. government influence to what might happen with foreign government influence can be a useful approach.

For example, China has publicly stated that they wish to be world leaders in artificial intelligence, semiconductors, and fast mobile networks, and they are already leaders in solar cells. Given that China is a serious competitor to the United States and has been taking technology from us for decades should resonate with patriotic people.

Interestingly, China is investing heavily in U.S. startups in artificial intelligence, autonomous vehicles, augmented/virtual reality, robotics, and blockchain technology [36], and by so doing are acquiring the technological know-how developed in this country. They went from producing 1 percent of the world's solar panels in 2001 to 66 percent in 2019. They are also dominating the world's wind turbine market and electric vehicle production [37]. It seems to me this would be hard for conservatives to ignore.

Those conservative states that are resistant to global warming action are in fact the ones to be most affected by its consequences [38], such as lower crop yields, coastal damage, and heat stress focused in Florida and other states along the Southeast and Gulf Coast. And this will directly affect the residents and their families, something we should all be sensitive to.

On the positive side, 225 of the world's largest corporations believe there will be $2.1 trillion in business opportunities due to global warming [39]. And, this will occur soon. Better to get in on a piece of this action early before it all flies overseas.

Fortunately, cracks are appearing in the conservative veneer. The American Conservation Coalition (ACC) states, "ACC believes that economic and environmental success can go hand in hand, and conservatives should champion this message and take a seat at the table in discussions concerning conservation, clean energy, sportsmen's rights, agriculture, climate, and much more [40]". Plus a Pew Research Center survey showed that young Republicans are more concerned than their elders by a two to one margin [41].

Politicians

You walk into Congressman Lamar Smith's office, retired but formally the chair of the House Science Committee, a position that you would think would make him sympathetic to global warming facts and want to take mitigating action. You would think. But you would be wrong. Congressman Smith has repeatedly attacked global warming science, going so far as to tweet out an article denying climate change and loaded with misinformation, published by the Breitbart News Network [42].

Why? Look no further than his donors and his constituency. As of 2015 he received contributions of $600,000 from oil and gas companies, more than any other industry [43]. And these gas companies are very much opposed to global warming science, as they view it as an assault on their business model and their ability to make money. Further, he represented the Twenty-First Congressional District in Texas, a conservative bastion.

It's not a question of whether you can convince him or not, it's just not in his best interest to agree and very much in his interest to disagree. You're wasting your time and his, so just walk away and fight another battle.

But don't blame Congressman Smith, or any other politicians who oppose global warming facts and actions. You need to put yourself in their shoes and understand what motivates them, what drives their actions. You might guess it's what's best for the country that motivates them, and to some extent it does. But they can't do what they believe is best if they don't get reelected, which is probably their greatest motivation. So their voting constituency, donors, bringing money and jobs into their district, making deals on the floor; this is what motivate them and this is what you must address.

Vanity, power, and money are also motivators, as they are for all of us, but given the power-laden high visibility of the job, politics tends to attract those with outsized egos who reside on the narcissistic side of the personality spectrum [44].

And if those are not enough obstacles for you, politicians are overwhelmed by lobbyists, supplicants, constituents, anyone who has an opinion, all clamoring for their attention with a bewildering crisscrossing of requests, truths, lies, and emotions. It might as well be the Tower of Babel for all the good it does to try to untangle this continuous bombardment of drivel. And here you come, with yet another appeal. Good luck with that.

If you want to be successful at this then do exactly what the politicians themselves do. They cut through this tsunami of conflicts with a simple criterion: "What is best for me?" In particular:

• How will this affect the voters in my district?

• How will this bring jobs into my district?

• How does this get me further along in my career?

• How will this generate favorable publicity for me?

• How much money will you donate?

So we must frame our arguments to address these items and no others. Fail at this and we might as well bang our heads on a sharp picket fence.

For instance, appealing to voters' financial needs, such as job creation from solar manufacturing plants, would win them over with no thought to global warming. Or appealing to their parental instincts to protect their children from future calamities is another good approach. This will push them to vote for global warming solutions.

Robb Willer and Jan Gerrit Voelkel discovered that one technique used by politicians is to frame political issues in terms that each side of the political spectrum understands [45]. For liberals, global warming issues "can be based on principles of economic justice, fairness and compassion," whereas for conservatives it should be "based on respect for the values and traditions that were handed down to us: hard work, loyalty to our country, and the freedom to forge your own path."

Donations always get you through the door. Big donations get them to listen. Really big donations get them to take action. Sorry, but that's just how it is. That's not to say there aren't other ways, but this one is a killer app. So getting organizations to pool their money and lobby congressional members who are on the fence is an important approach. Remember, Political Action Committees (PACs) and SuperPACs cut both ways. They can be used to prop up fossil energy agendas, but they can also be used to push legislators on global warming actions. I am opposed to PAC and SuperPACs, but as long as they exist, we might as well use them if we can.

Of course we do have a vote. I admit a vote isn't what it once was or what our founding fathers designed it to be—not with PACs, SuperPACs, and mega-donors pushing billions of dollars into elections and greatly affecting the outcomes. Nonetheless, a vote is still a vote and it's ours to use.

I say politicians, but I also mean anyone challenging an incumbent for their seat. Here we need to make a Machiavellian decision, as to where our resources are best used. That is, who is more likely to win and help our cause?

And we need to choose our politicians carefully. I don't think it's a coincidence that fossil energy companies target conservative state politicians, since the fossil energy interests align with the voters' interests, which makes it easier to convince the politicians to do their bidding. The politicians can take the lobbyists' money and still be true to their voters. We need to do the same. And while conservative states are starting to see the light, until we can count on their supporting pro–global warming politicians, it may be best to skip over them so as not to waste money and focus on the battles on the margins where there is no conclusive position and ones in which we can sway the results.

Why target politicians in the first place if convincing them is so difficult? Because they make the laws and control a $4.4 trillion budget. It's the Golden Rule at its finest—those with the gold make the rules. No wonder 11,102 registered lobbyists have their offices in D.C. and spent an incredible $1.75 billion in 2019 trying to affect politicians' votes [46]. And while there are other avenues to pursue, getting support from Congress will open up a floodgate of federal actions that could not be achieved any other way.

But be aware that this is a double-edged sword. While getting federal action has advantages, it conflicts with conservatives' desires to limit government involvement, thus making it harder to convince them. As with all things in life, no solution is perfect and a balance must always be struck.

"As an Aside, just when I thought I had it all figured out, along comes the culture wars. Rivian Automotive wants to build an electric truck plant outside of Atlanta, spending $5 billion and creating 7,500 jobs, which would be the "largest economic development plant in Georgia's history [47]". So you have an economic boon for the local area that doubles as a major push to solve our global warming problems. Can't miss, right? Not right. Never underestimate the power of power. Or at least the grab for power. David Perdue, in his bid for the Georgia governor's Republican nomination, tried to derail the project, claiming Rivian is a woke corporation and should be sent packing its bags back to California. And Perdue actually started to gain traction among some of the state's voters".

"Above, I went on about the wonders of job creation to move a politician to our side. And now, Perdue demolished my argument in a single, absurd gambit for the nomination. Anyway, Perdue lost by a landslide, so maybe he is the exception that proves the rule and my approach works after all. Whew. Close call".

The Unconvincibles

Why waste time trying to convince the Unconvincibles, those people so cemented in their views that nothing will budge them? Because there may be onlookers listening and judging. Use the Unconvincibles as a platform to get to your real audience and play to the onlookers' Metabeliefs. Know the crowd you are working. It may not be the one you are face-to-face with but the onlookers. Adjust accordingly.

And keep in mind we are selling ourselves as much as our position, maybe more so. So we must come across as likeable, honest, and knowledgeable, and if we can pull it off without being insincere, as one of them—that is, as part of their group, clan, tribe, whatever, so they can identify with us.

Those Already Convinced

You would think preaching to the choir would have little benefit and not be worth our time—after all, we want to get the biggest bang for our buck, and if they are already convinced, what do we have to gain? But according to the *Investor's Business Daily*, studies from the University of Michigan and Cornell University showed that those who are highly concerned about global warming are least likely to take individual action but rather will rely on government involvement [48].

Maybe we need to nudge them toward action and away from self-absorption.

The Influencers

It fascinates me that religious leaders' messages get drowned out by cultural and political influencers. Just think how powerful these influencers are when they can get folks to ignore their religious beliefs, which oftentimes are strongly held.

Clearly this points out the obvious need to convince the influencers first, which will open a much easier path for their followers. Commercials work this angle all the time. Why do you think they have celebrities, rather than experts, hawking their products? Would you rather listen to a grease monkey or a smooth-talking, handsome Matthew McConaughey talking about Mercury Lincolns?

I sense my readers' unease is surfacing again.

"How can we influence the influencers when we are just normal Joes and Janes?"

First of all, we are not alone, as we are a part of an almost Borg-like collective that gives us strength in numbers. As a voting bloc, religious beliefs, political beliefs, team affiliations, any group with similar interests, whatever; we have many allies. The challenge is to energize our many doppelgangers into action.

Second, we don't need to get the Matthew McConaugheys of the world on board, as nice as that might be, but rather work the grass roots with the local influencers in our congregation, at work, our bowling team, and the other groups we belong to.

Psychopaths

What do we do about psychopaths? Yes, I am serious. Some of the most successful executives are psychopaths. Not the kind you see on TV or read about. Not the ones who lack inhibitions and get off on violence. That's not what I mean. But rather I'm talking about those people who have little or no empathy toward others, are very bright very motivated, believe rules are not meant for them, and are in substantial positions of power. These folks are quite capable of violence but refrain from it as it will not further their ambitions.

It is their power that makes it necessary to deal with them and their lack of empathy that makes it difficult to do so.

They are influential and we will come across them sooner or later. You probably already have. The good news is that they have Metabeliefs and aspirations just like the rest of us, and we need to address those. Thus, motivating them should include their outsized ego and how others view them. So stroking their ego is the way to go here. If we want to get them to follow the global warming path, we must show them how important it will make them and how others will admire them for it.

I don't confine myself to just psychopaths but rather to all mental disorders who share a lack of empathy but are not debilitating. This includes narcissists, who, in addition to a lack of empathy, also have an exaggerated sense of their importance, a sense of entitlement, a constant need for admiration, and will exploit others.

Machiavellianism is another disorder that shares the traits of psychopaths and narcissists, with an added touch of duplicity.

As a group, these folks are predictable. In their minds, the universe revolves around them, which is the key to gaining their concurrence. We don't confront them or try to overtly control them. Instead, we use their weight against them. If they need to be the center of attention, then we put them in the center and let them take credit for the direction we want them to take.

Such people are more common than you might think. A study of 261 professionals found nearly one-quarter exhibited psychopathic traits [49]. That's surprisingly large. So that means if I am with three other people and they are not psychopaths, then I must be. Well, let's not go there.

But Sometimes Facts Do Work

Maybe not often, but there are times when facts turn the trick and get our target off the fence and squarely into our camp. And despite this chapter's theme, we should be prepared to grab this low-hanging fruit however infrequently it may occur.

Whom would these enlightened targets be? There are a number of possibilities. Those for whom global warming issues do not conflict with their most important Metabeliefs, those in the process of changing their Metabeliefs, the curious, and those who are well-traveled.

"When people are exposed to a more diverse group of friends, their brains are forced to process complex and unexpected information. The more people do this, the better they become at producing complex and unexpected information themselves[50]*"*. So, folks who travel, have a diverse set of friends, are familiar with multiple cultures, can move out of their comfort zone easier than others—these are all targets. This is not foolproof—nothing in this chapter is meant to be—but it puts the odds in our favor.

People who are curious about science are also good targets as they are more willing to accept positions that conflict with their political beliefs than are those not as scientifically curious [51].

People who care about and consider the future are to be embraced as well. In contrast, people who live from one day to the next and only care about the present should be avoided. Also avoid those yearning for the nostalgic days of yesteryear, living their lives in the past. Leave them to their dreams. They can't help us.

YECA suggests that Generation X care more about global warming than their parents, so focusing on them would be ideal. These are the youngsters born between 1965 to 1980 with an age range, as of 2022, of 23 to 36 years old. Those just out of their teens but not yet middle-aged. In other words, the younger generation will be more interested in learning than their parents.

And don't forget teachers. Middle and high school teachers provide a direct conduit to students who are just learning about global warming and whose minds may be more open than adults, though certainly, their Metabeliefs are likely well formed by now. In some sense, the teachers are our farm club, producing the players we will need in the future. But teachers need to understand the science before they can teach it. Unfortunately, it is not as one would hope. Three of five surveyed teachers were unaware of the near-total consensus among scientists that global warming is caused by us [52]. One-third of the teachers said global warming was not caused by us but was a natural occurrence. Clearly, this is an important group who need the right facts and quite likely would be more open to it than most, given their dedication to their teaching profession.

As an example, a waitress, who also teaches autistic kids, told me she was concerned that vaccines might be the cause of her kids' autism. Note that she was concerned but not convinced. After a few minutes of conversation, it was clear she just didn't know and was looking for answers. This young woman is an ideal target, one with an open mind, looking for the truth. I admire her, as she did not fall for the false news, lies, and what have you on this emotional topic. This is the type of person where the facts work if they are provided in a manner that doesn't conflict with any strongly held beliefs and in a non-confrontational manner.

Ignorance is our friend because it can be easily corrected if someone is curious. On the flip side, arrogance, its evil twin, becomes a tortuously uphill climb when it is sustained by a person's Metabeliefs.

There are a number of excellent websites that provide answers to typical global warming questions, which you can find by googling "global warming hoax," "global warming facts," and "climate change skeptics." One good one is from the

Skeptical Science site "Global Warming and Climate Change Myths [53]".

An example of a good one is Mark Maslin's "Five Climate Change Science Misconceptions—Debunked [54]", which I have reworded here.

<u>Global warming is just part of the natural cycle.</u> **False**.

It is true that the climate has always changed but never this fast and to this degree. The below figure shows how CO_2 levels have bounced around for the past 400,000 years before rocketing up immediately after the Industrial Revolution [55]. Not only are the CO_2 levels unprecedented, but more importantly, the change, on a geological scale, has occurred faster than a smash-and-grab thief, dwarfing previous rates of change, which provides precious little time to adapt.

The final nail in the coffin is the next figure, which shows how our increasing temperature goes in lockstep with CO_2,[56]. So, we are pouring historic amounts of CO_2 into the air, which is increasing our temperature to dangerously high levels never seen before.

Carbon Dioxide Levels Have Skyrocketed In Modern Times

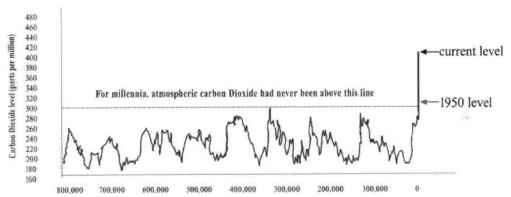

Global warming is due to sunspots or to an increase in the sun's thermal output.
False and False.

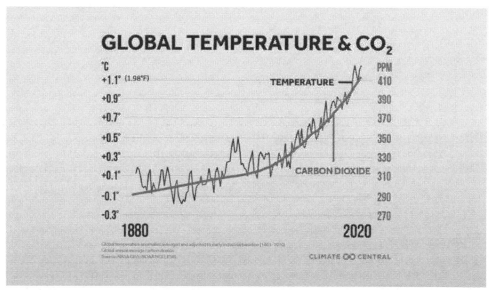

The figure below shows the global temperature increase from 1880 to about 2018 and the corresponding sun's thermal output [57]. From 1960, the global temperature has increased, whereas the sun's output has stayed constant.

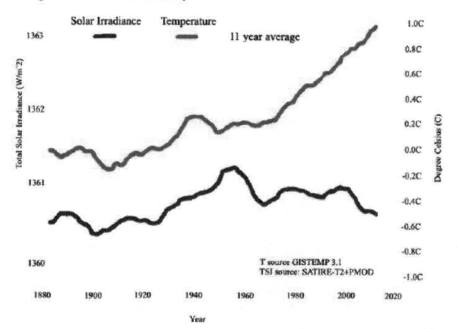

Global warming is due to galactic cosmic rays. **False**

Someone please tell me how this one got started. Cosmic rays? Really? Cosmic rays are high-energy particles that mostly get deflected away from the earth due to its magnetic field (though less so at the equator) and also our atmosphere hinders their descent. But in any event, they do not affect clouds, and while energetic, there are not enough of them to heat the earth. They can have an effect on aircraft crews and passengers who fly at high altitudes and astronauts in space, but that's about it.

CO_2 is a small part of the atmosphere and therefore cannot have a large heating affect. **False**.

The first part is quite true and at first blush the conclusion seems like a reasonable objection. There are only 420 ppm of CO_2 molecules, where "ppm" stands for parts per million. That is, there are 420 CO_2 molecules swimming in a sea with 9,999,580 air molecules. Looking for them would be like trying to find a pebble in the bottom of an Olympic-size swimming pool. But remember, even small things can pack a big punch. For example, only 20 ppm of arsenic will kill a person,[iii] so at least in this simple-minded analogy, the harm of 420 ppm of CO_2 does not seem outlandish.

Also, consider the extreme ranges. If our atmosphere contained nothing but CO_2 (one million ppm), the sun would relentlessly heat the earth, boiling off all of our oceans and vaporizing most materials (including us). Venus is an example of a runaway greenhouse effect, and lead would melt on its surface.

On the other extreme, if there were no CO_2 in our air, then more heat would escape the earth, and the temperature would plummet, freezing our oceans. Not as dramatic as a melting lead, but for us Homo sapiens, the effect is the same—no us.

Two hundred and eighty ppm is the Goldilocks amount, neither too hot nor too cold but just right for us, thank God (literally?). In fact, that is exactly what it was before the Industrial Revolution began spewing CO_2 into our air.

Now consider the current 420 ppm. It is north of the perfect 280 ppm, which puts it on its way to Venus-like conditions, so it must heat up our planet and everything on it. And the higher CO_2 climbs, the hotter our planet will get; an unpleasant trajectory to think about.

Scientists manipulate all data sets to show a warming trend. **False.**

This is a typical conspiracy theory without foundation, used to either generate

---

[iii] Compared to our body weight.

internet clicks or promote an unsustainable and illogical position. It would require the cooperative collusion of hundreds of thousands of people over the entire world. Come on, we can't even get opposing sports fans to talk to each other and you expect thousands of diverse people to agree to fool their friends, relatives, and everyone else. And do it with perfect coordination rivaling a well-orchestrated ballet. Like most conspiracy theories, this is as false as saying the moon is made of green cheese.

Connecting The Dots

1. We must convince others that global warming is occurring and action to reverse it is essential to our well-being. Sure, we may understand it ourselves, but we need everyone to pull together if we are going to have a chance of overcoming this calamity. We can't do it by ourselves.

2. Unfortunately, it does no good to confront others with an array of facts, however true, if it conflicts with their deeply held Metabeliefs. Such an approach is futile. It will only force them to retreat into their ancient instincts of fight or flight, and any chance of convincing them is as futile as offering aspirin to a corpse.

3. The first order of business is to determine our targets' stripes. Where do they fall on the spectrum of global warming beliefs? Are they in our camp, on the fence, or hopelessly against any action to reverse global warming? We need to nudge the first, convince the second, and avoid the third.

4. And let's not threaten or cause folks to fear global warming. That route will just force them into a fetal position, frozen in fear and unable to act. Rather, we need to show that global warming issues are tractable and, acting together, we can solve them. Difficult? Yes. Impossible? No.

5. Let's be upfront and honest with those we talk to. Stretching the truth, no matter our intentions, will only push people away when they inevitably discover we have been dealing from the bottom of the deck. Better to play the long game with honesty even if we miss a few near-term goals. Plus, it's good for the soul.

6. Similarly, we should treat our opponents with respect and listen to their arguments. Again, we are playing the long game by winning them over with trust, not pushing them away with unwinnable confrontations. Besides, we might learn something.

7. Compromising is a great tool that will win many converts. We want folks to help solve global warming. We want nothing else. So giving in on other points, if

we score our main one, is the only way to be successful.

8. And let's not kid ourselves. Time is no friend of ours, so we must move quickly and decisively. According to the IPCC, we must eliminate all greenhouse gases in less than 30 years. That's a blink of an eye.

9. Sooner or later there will be a serious, well-publicized disaster from global warming. Unfortunately, this is a certainty. In fact, we are likely to have a number of them marching toward us and getting more extreme and more frequent. So when the inevitable occurs, we should be ready to pounce with our campaign to get our message across. Nothing like a disaster to focus peoples' attention.

10. And let's not get complacent. No matter how many disasters and how well our campaign goes, people will always find a way to rationalize it away so they remain safe in the womb of their false beliefs. We need to recognize this and deal with it.

11. We have many tools to convince others. Once we know where they stand, and a little about their beliefs, we can focus our approach to align with their Metabeliefs. There is no other way. But if done well, it will work.

12. Occasionally, facts will convince some people. So we need to also be prepared for this.

13. The actions in this chapter are not meant to be perfect or to work all of the time. It doesn't' need to. It just needs to work some of the time to get enough people to take action. But what if we fail? Then sayonara. Which is another way of saying we *must* make it work. We have no choice. Or as NASA said during the Apollo 13 explosion, "Failure is not an option."

Answers to "What Are Their Global Warming Beliefs?"

Person	Will Likely Support Action	Will Likely Not Support Action	Cannot Be Determined
Black Female	X		
White Protestant		X	
Reads *Wall Street Journal*			X
White, Attended College, Not Religious	X		
White and Gay	X		
White and Jewish	X		
Has a College Degree	X		

(Table) cont.....

Person	Will Likely Support Action	Will Likely Not Support Action	Cannot Be Determined
Asian and Married	X		
Asian Protestant			X
Black Male under 31	X		
Older White Male		X	
Unmarried Female	X		
White Male Did Not Attend College and Not Religious			X
Watches Fox News		X	
Lives in New York	X		
Lives in West Virginia		X	
Wealthy Religious Hispanic		X	
Pope Francis	X		

Afterthought: Selling Refrigerators To Eskimos

The old sales adage is a good salesman can sell refrigerators to Eskimos, which is utter nonsense of course. Eskimos' official currency is ice coins so they'd be selling a counterfeiting machine, which will get them thrown into the slammer which is also made of ice. It gets a bit monotonous in the frozen north.

Well, maybe I exaggerated but the point is a good salesman (woman) should sell what the customer needs not what the salesman wants.

Maybe a good salesman is one who can sell honesty to politicians. Oops, looks like we're back to selling refrigerators to Eskimos. That's the last thing a politician needs or wants. Though you must understand who the customer is—in this instance it's not the politician but the voter. This is what voters desperately need.

But I doubt it will ever happen. Remember, you can lead a polar bear to water, but you can't put a bathing suit on him. Come to think of it, I'd like to see you try leading a polar bear anywhere. All the best to you. The point is, even superb sales folks can't do the impossible, so let's forget about upgrading our politicians.

This leaves us with us. Remember us? We're the people the Constitution gave complete power to choose our leaders. I know it feels like a long time ago in a galaxy far, far away, but it did happen and we do have that power. Trust me on this one. I wouldn't lie. You can look it up. The Constitution clearly states "The

Illuminati hereby grants people the illusion to choose their own leaders, when in fact it is done in smoke-filled rooms." Don't worry. With all that smoke, the powers that will be soon die off from lung cancer and we'll be back in the driver's seat.

So what voters really need is to pick honest, competent leaders, so maybe we need a salesman (woman) to sell us the abilities to do so. You know, the ability to see through the trickery and be able to convince ourselves and others as to who we should vote for based on what is best for our country and not what is best for the moneyed influencers.

Now, I'd pay real ice coins for that.

References

[1]   Shermer M. For the Love of Science. Sci Am 2019.
        [PMID: 29257816]

[2]   Wolchover N. Are Flat-Earthers Serious? Live Science 2012.

[3]   Hayhoe K. When facts are not enough. Science 2018; 360(6392): 943.
        [http://dx.doi.org/10.1126/science.aau2565] [PMID: 29853662]

[4]   Funk C, Kennedy B. Public Views on Climate Change and Climate Scientists. Pew Research Center
        2016.

[5]   Lucchdesi N. Misrepresentations of Climate Data Come Down to Political Loyalty 2018.

[6]   Otto S. A Plan to Defend against the War on Science. Sci Am 2016; 9.

[7]   Chinoy S. Quiz: Let Us Predict Whether You're a Democrat or a Republican New York Times 2019.

[8]   Marlon J, *et al.* Yale Climate Opinion Maps 2018.

[9]   Troy D. Don't bother waiting for conservatives to come around on climate change. 2016.

[10]   Wadman M, You J. The vaccine wars. Science 2017; 356(6336): 364-5.
        [http://dx.doi.org/10.1126/science.356.6336.364] [PMID: 28450592]

[11]   Moyer M. W., "Why We Believe in Conspiracy Theories,". Sci Am 2019.

[12]   The Fukushima Daiichi Nuclear Disaster. Wikipedia.

[13]   Ray M. Paul Cairney: Politics & Public Policy.

[14]   Farley R, Roberson L. Trump's False Claims about Rep. Ilhan Omar. 2019.

[15]   Montoya-Galvex C. Nearly Half of Young Americans Say Climate Change Is a "Crisis" Requiring
        'Urgent Action,' 2019.

[16]   IPCC, "2018: Summary for Policymakers. In global warming of 1.5oC…", V., Mason-Delmontte, *et
        al.*

[17]   Barnard M. How to Talk to Conservatives about. Clim Change 2018; 13.

[18]   Well K. Oregon statehouse shut down after lawmakers team up with right-wing militias. The Daily
        Beast 2019.

[19]   Wessel L. False: Vaccination Can Cause Autism. Science 2017; 356(6336): 28.
        [PMID: 28450594]

[20]   Van Bavel J. The Problem Is Us. MIT Technology Review.

[21]   Zukerman W. The Trouble with Science: We just won't accept unintuitive facts 2014.

[22]   Global Warming, God and the 'End Times,' Yale Program on Climate Change Communication 2018.

[23]   Encyclical Cal Letter Laudato Si' of the holy father francis on care for our common home.

[24]   Church Backs Climate Action. Science 2018; 361(6398): 13.

[25]   "Climate Change Statements from World Religions," Forum on Religion and Ecology at Yale.

[26]   Goldhill O. The Green New Deal Isn't Socialist, It's biblical, argue evangelical environmentalists.

[27]   Sachedva S. Religious Identity, Beliefs, and Views about. Clim Change 2016.

[28]   Ibid.

[29]   Meinecke S. God and the Earth: Evangelical Take on Climate Change 2019.

[30]   Available from: https://www.yecaction.org/

[31]   Deaton J. Racial Resentment May Be Fueling Climate Denial, Study Says. Nexus Media 2018.

[32]   Kahan D. Fixing the communications failure. Nature 2010; 463(7279): 296-7.
[http://dx.doi.org/10.1038/463296a] [PMID: 20090734]

[33]   Meinecke S. God and the Earth: Evangelical Take on Climate Change 2019.

[34]   Corso J. Report Shows Texas 2nd in Nation for Clean Energy Jobs. San Antonio Business Journal 2019. Available from: https://www.bizjournals.com/sanantonio/news/2019/03/14/report-shows-texas-2nd-in-nation-for-clean-energy.html/

[35]   Funk C, Kennedy B. Public Views on Climate Change and Climate Scientists. Pew Research Center 2016.

[36]   Brown M, Singh P. China's technology transfer strategy: how chinese investments in emerging technology enable a strategic competitor to access the crown jewels of US Innovation. Defense Innovation Unit Experimental 2018.

[37]   Helveston J, Nahm J. China's key role in scaling low-carbon energy technologies. Science 2019; 366(6467): 794-6.
[http://dx.doi.org/10.1126/science.aaz1014] [PMID: 31727813]

[38]   Climate Harms Red States. Science 2019; 363(6426): 436.

[39]   Palmer B. Companies See Climate Change Hitting Their Bottom Lines in the Next 5 Years. New York Times 2019.

[40]   What Is the American Conservation Coalition?

[41]   Brady J. 'Light Years' ahead of their elders, young republicans push GOP on climate change. NPR 2020.

[42]   Raymond L. The House Science Committee Tweets Are an Embarrassment to Science. 2019.

[43]   Lamar Smith 2019. Available from: https://en.wikipedia.org/wiki/Lamar_Smith

[44]   Hill RW, Yousey GP. Adaptive and maladaptive narcissism among university faculty, clergy, politicians, and librarians. Curr Psychol Res Rev 1998; 17(2-3): 163-9.
[http://dx.doi.org/10.1007/s12144-998-1003-x]

[45]   See De Witte M. The Power of Dressing Progressive Economic Policies in Conservative Clothes. Stanford Business 2019.

[46]   Lobbying Database 2019.

[47]   Seitz-Wald A. Georgia electrical vehicle factory becomes kemp, perdue campaign battle. Politics News 2022.

[48]   Climate-change true believers are least likely to change their own behavior, study finds Investor's Business 2018.

[49]   Waugh R. Six giveaway signs that you might actually be a psychopath. Yahoo News 2018.

[50]   Van de Vyver J, Crisp R. Crossing Divides: The friends who are good for your brain. BBC News 2019.

[51]   Scientific Curiosity verses Polarization. Science 2017; 355(6326): 17.

[52]   Procopiou C. Overwhelming percentage of science teachers confused about climate change. Newsweek 2016.

[53]   Global Warming and Climate Change Myths Available from: https://skepticalscience.com/argument.php/

[54]   Maslin M. Five Climate Change Science Misconceptions—Debunked 2019.

[55]   Global Climate Change. Available from: https://climate.nasa.gov/evidence/

[56]   Climate Central. Available from: https://ccimgs-2021.s3.amazonaws.com/2021CO2Peak/2021CO2 Peak_Temps_en_title_lg.jpg/

[57]   Graphic: Temperature *vs* Solar Activity, NASA, Global Climate Change, Vital Signs of the Planet. 2019.

CHAPTER 3

Finding the Truth Amidst a Sea of Lies

"A lie by any other name is still a lie"

—*appropriated from Shakespeare*

Hucksters pedaling deceit and half-truths are as old as humanity, starting when Adam and Eve were ejected from Paradise because they could not recognize the serpent's lies. I guess when you live in Paradise, there is little reason to be on your guard.

Their fall should not be surprising since we are built to survive, not to uncover the truth[1]. If the truth gets in the way, it becomes nothing more than collateral damage. Think about it for a moment. If strength can't overcome the alpha male, then deception is the only way left. And it usually works quite well.

So why then should we care about the truth? What good does it do us? Might it even be better to embrace lies so as to advance our cause? Fight fire with fire, so perhaps we should attack their lies with our own lies, their falsehoods with our own better-crafted falsehoods. Meet them head-on with a stronger narrative neglecting all facts that don't support our position. Why not surrender and sign the devil's pact if it gets us what we want?

There are lots of reasons. First, our integrity and the legacy we leave to our children and grandchildren should be important to all of us. Lies tarnish both. Do we want to be remembered like Lance Armstrong, who acknowledged he lied "10,000 times" during his infamous doping scandal [2]?

Most importantly, lies drastically hurt society by forcing poor, even catastrophic, decisions. How can the outcome of any decision be good if it is based on lies? The tobacco industry lied about the hazards of smoking, which now leads to 480,000 deaths annually [3]. Not good.

And to be a bit melodramatic, the truth will always win. Not at first maybe, but eventually. It took hundreds of years for people to accept the earth revolves around the sun, and for giants like Copernicus, Galileo, and Kepler to achieve it. But eventually, the truth won out.

All lies die—they are no match for reality. But during their lifetime they can do a boatload of harm. Just look at the global warming damage already incurred because fossil energy companies want to sustain their profits as they put up obstacles at every turn. They have even been accused of misleading their own investors on the global warming costs to their bottom line [4], [5]. And now, thanks to their persistence and our acquiescence, we are undergoing record heat waves across the globe, crumbling arctic ice, leading to dangerous sea level rises and mass extinctions seen only five other times throughout earth's 4.5-billion-year history.

As Adam and Eve found, we must defend ourselves from lies. And while we can't go back to Paradise, we can protect the second best—our world as it is today—from global warming's ravages.

CHAPTER CONTENTS

A Balance of Facts

I will experimentally prove the earth is flat, and I will do it with scientific rigor. Here's what I did. I used an app on my smartphone that, through the phone's camera, measures the vertical distance above or below a target point.

Next, I planted a pole in the ground and focused the app on its top. I recorded my distance from the pole and the distance from the top of the pole to the ground. I then moved it a few feet and did it again. I continued to move farther away from my starting point and stopped occasionally to record my distance and the vertical height of the ground from the top of the pole.

If the earth is round, the vertical measurements should trend downward from my starting point as it would if I were walking on the surface of a large ball. And if the earth is flat, the measurements should be more or less constant as if I were walking on a flat board.

The following figure shows the results. The vertical distance from the top of the pole is plotted (on the y-axis) versus the distance from the starting point (on the x-axis). There is clearly scatter in the data, but for the most part, the data is constant, neither rising nor falling, and varies by only a few hundredths of an inch from a flat surface. I can therefore say with certainty that, within the limits of my data, the earth is flat.

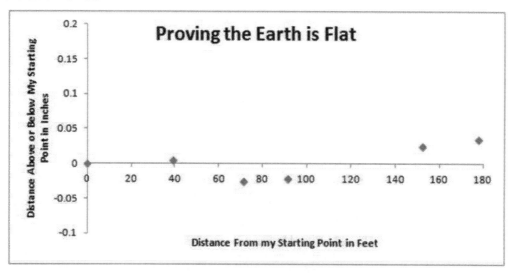

I can hear your first objection: *"Your data must be wrong. You made a mistake."*

Pshaw. My data is unassailable. I provided my methods, its accuracy, and the plot is the raw data. You have everything you need to assess it. And if you tried to duplicate it, the hallmark of good science, you will get the same results. Of this, I am sure. You, my imaginary friend, are wasting your time trying to refute it.

"But you only went 180 feet. What about after that?"

That's an unwarranted and spurious objection. How can I possibly know, and how can you possibly expect me to know, what the data should be outside of my measurements? I can't report data I never took. Shame on you, asking for the impossible.

And I am sure I can get this published in a journal, maybe not the most respectable one but a journal nonetheless.

I'll return to this flat earth dilemma in a moment. Now it's time to play a card game.

You are dealt a pair of aces in a Texas Hold'em high-stakes game, and you know you have an 80 percent chance of winning. So you bet heavily on each turn. The *flop*, or the first three community cards that all players can use, are a 3♥, 9♠, and 6♣. You are on a roll, so you raise the bet. The next card, called the *turn*, is a 9♦. Getting a bit dicey, but you still hold the advantage. The final card, the *river*, is a king♦. You call and your opponent shows her hand, in the following figure.

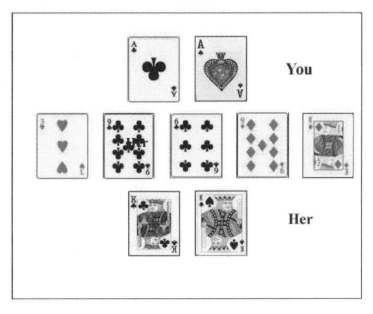

What started out with so much promise just crashed and burned? Your opponent wins as she has a full house (three kings and two nines) to your two pairs (aces and nines).

So what went wrong and what happened to your near-certain chance of winning? Nothing. You played expertly and you used the incomplete information you had available. Eight out of ten times you would win and come away with a lot of money.

Like all things in life, we never have complete knowledge, and we must make decisions based on what we know and what we can infer. It's never perfect and we will sometimes be wrong, but if we are true to the facts we possess, we will be mostly right. The card game was played as well as it could be, but sometimes shit happens. That doesn't invalidate the case for facts; it simply acknowledges the imperfect nature of what we can possibly know.

Now back to my flat earth measurements. My data is irrefutable, but you must, and let me emphasize the word *must*, place it side by side to all the other facts that support a round world. The round earth is based on countless observations going back to the ancient Greeks. They knew the earth was round by watching ships sailing toward them. First they saw the top of the mast, then the rest of the mast, and then the ship itself. This never failed. Only a globe could account for it. And photos taken from the International Space Station always show the world to be round. I could go on and on. The point is, a single contrary fact, when compared to the almost infinite facts arrayed against it, should be washed away like a sand castle at high tide.

If you learn nothing else from this book, learn the Balance of Facts Principle:

You cannot take a single piece of evidence that supports your position, no matter how appealing it might seem, and ignore the universe of data and facts that argue against it.

And its corollary:

Do not let a single fact derail you. It is the balance of facts that count.

And that is why my flat earth data, as good as it is, should be immediately dismissed; swept into the dustbin and forever forgotten.

Here's an interesting example. The IPCC stated that 97 percent of all scientists agree that global warming is happening and we are responsible. A blogger in response posted this clever sound bite: *"Well, would you get on an airplane that was only 97 percent safe?"*

The blogger intentionally misrepresented the true situation since it implies we have a choice of boarding the plane or staying on the ground. With global warming we have no such choice. We can't get off the planet. So the correct analogy should be *"Would you get on a plane that was 97 percent safe or one that was only 3 percent safe?"* Now the answer is clear. And it comes back to balancing all the facts, not just a select few. Like the blindfolded Lady of Justice with a scale in one hand, there will always be conflicting facts, but the only important thing is which way the scale moves. And with global warming, her scale would be so over weighted she would tilt and fall.

Unfortunately, there are plenty of places to get evidence to support any and all positions. As Hester Hill writes,

"The world's 6,500 peer reviewed scientific journals disgorge 4 scientific papers every minute of every day. That's more than 2 million good, bad, and mostly mediocre papers every year. It should be no surprise then that a study can be found to back up virtually any point of view."[6]

Plus there are predatory journals out there as well with no peer review, with fake editors, designed to support a biased position with no intent of honesty or balance.

As with the card game, there will always be uncertainty. Even Newton's law of gravity was overturned by Einstein's general theory of relativity. But uncertainty compared to what? You need to judge the uncertainty of global warming to that of its opposite, no global warming. Uncertainty does not imply fault. It is simply an affirmation of reality. It is always the balance of facts that speak the truth.

Fact-checking

'If there's something strange on the internet,
Who you gonna call?
Fact checkers!'

Some internet outlets slime us with misdirection and lies to legitimize whatever false ideas they are peddling. And it is surprisingly widespread as one study concluded that 86 percent of people believe they have been duped by fake news[7].

Why would anyone do that? Money is a big reason. The more clicks on their site, the more money they make. Or maybe to spread a political message that can't survive a rigorous inspection, so an end around is made hoping no one notices. These types of people heartily embrace "the ends justify the means" rationale. It's all the rage these days among the truly incompetent. Or maybe just plain everyday anarchists who wish to tear down rather than build. And who knows what else.

But there is also a Camelot element to the internet, a part that exposes lies the way a proton pack dispatches ghosts.

A good one is from *Skeptical Science*, a premier global warming myth buster. Here's one example from their site:

<u>Myth</u>—"Humans are too insignificant to affect global climate."

<u>Busted</u>—"Humans are small but powerful, and human CO_2 emissions are causing global warming. Atmospheric CO_2 levels are rising by 15 gigatonnes per year. Humans are emitting 26 gigatonnes of CO_2 into the atmosphere. Humans are dramatically altering the composition of our climate."

There are almost two hundred such myths busted.

Other fact-checking sites are given in the following table. It doesn't include all of them, but it is an excellent place to start.

Fact-Checking Sites.

Site Name	URL Link	Description
Climate Feedback	climatefeedback.org/	Checks media global warming coverage
Desmog	https://www.desmogblog.com/	Fixes global warming disinformation
Fact Check	factcheck.org/	Fact-checks politicians' statements
Flack Check	flackcheck.org	Part of Fact Check, it mostly reviews videos
Hoax Slayer	https://www.hoax-slayer.com/	Unravels urban legends, myths, etc.
Media Bias Fact Check	https://mediabiasfactcheck.com/	Evaluates hundreds of media sites for accuracy and political leanings
National Center for Science Education	ncse.ngo/climate-change	Provides information on global warming for students and educators
News Guard	newsguardtech.com	Rates news sites and is appended to a browser
NPR	npr.org/sections/politics-fact-check	Fact-checks political statements
Open Secrets	OpenSecrets.org	Tracks political donations and their effect on elections and public policy
Politifact	Politifact.com	Fact-checks politicians' statements
Skeptical Science	skepticalscience.com	Debunks global warming myths
Snopes	Snopes.com	Unravels urban legends, myths, etc.
SourceWatch	SourceWatch.org	Covers corporate misinformation
Sunlight Foundation	https://sunlightfoundation.com/	Makes government transparent

(Table) cont.....

Site Name	URL Link	Description
Truth or Fiction	https://www.truthorfiction.com/	Unravels urban legends, myths, etc.
Washington Post	washingtonpost.com/news/fact-checker/	Fact-checks political issues

To Trust Or Not To Trust

Next up is a way to evaluate the honesty of news sites both on the internet and in print. For the younger generation, print refers to paper copies. "*Huh?*" Paper is made from trees. Nothing yet? Paper is something you can write on with a pen. "*A pen?*" Never mind.

I tabulated a number of news outlets by passing them through an outstanding site, Media Bias Fact Check (mediabiasfactcheck.com[i]). They rate news outlets on their political leaning and on how factual they are. The factual content goes from best to worst as Very High, High, Mostly Factual, Mixed, Low, and Very Low. Clearly, you want to shoot for Very High, High, or Mostly Factual in anything you read.

I catalogued the sites into the Good, the Bad, and the Ugly type of tables based on their factual reporting, with no regard to their political biases.

The first table consists of sites with Very High (VH), High (H), and Mostly Factual (MF) site ratings. This is the Gold Standard and this is where you should be getting your news from.

Gold Standard Table.

News Outlet	Political Bias	Factual Reporting
110 WINS AM	Least Biased	**H**
ABC News	Left of Center	**H**
American Conservative	Right of Center	**H**
Associated Press	Least Biased	**VH**
Atlantic	Left of Center	**H**
Axios	Left of Center	**H**
BBC	Least Biased	**H**
Bloomberg News	Left of Center	**MF**
Business Insider	Left of Center	**H**
Cato Institute	Right of Center	**H**
CBS News	Left of Center	**H**
Climate Feedback	Unknown	**VH**

---

[i] Media Bias Fact Checking was free when I started writing this book, but all good things must come to an end, I suppose. Still, you can get a $1/month subscription, so that's not too bad.

(Table) cont.....

Cosmopolitan	Left	MF
C-Span	Least Biased	VH
Current Affairs	Left	H
Daily Beast	Left	MF
Daily Signal	Right	MF
Economist	Least Biased	H
Esquire	Left	MF
Factcheck	Least Biased	VH
The Fiscal Times	Right of Center	H
Forbes	Right of Center	MF
Fortune	Right of Center	H
Forward Progressives	Left	H
Google News	Left of Center	H
Harvard Business Review	Least Biased	H
The Hill	Least Biased	MF
Intergovernmental Panel on Climate Change (IPCC)	Least Biased	VH[ii]
Jacobin	Left	H
Knoxville News Sentinel	Least Biased	H
Los Angeles Times	Left of Center	H
Market Watch	Right of Center	H
Marshall Project	Least Biased	H
Mashable	Left	MF
MIC	Left of Center	H
Mother Jones	Left of Center	H
National Center for Science Education	Unknown	VH
National Review	Right	MF
NBC News	Left of Center	H
New Republic	Left	H
New York Times	Left of Center	H
New Yorker	Left	H
News Channel 4	Least Biased	H
NPR	Left of Center	VH
PBS News Hour	Least Biased	VH
Pew Research	Least Biased	VH
Politico	Least Biased	H

(Table) cont.....

Politifact	Least Biased	H
Poynter Institute	Least Biased	H
ProPublica	Left of Center	H
Reuters	Least Biased	VH
Rolling Stone	Left	H
Second Nexus	Left	MF
Slate	Left	H
Time	Left of Center	H
TPM—Talking Points Memo	Left	MF
USA Today	Left of Center	H
Vanity Fair	Left	MF
VOX	Left	MF
Wall Street Journal	Right of Center	MF
Washington Post	Left of Center	H
The Week	Left	H
Weekly Standard	Right	H
Yahoo	Left of Center	H
Young Turks	Left	MF

Factual reporting legend: VH—Very High, the best factual rating H—High MF—Mostly Factual

Next, the Meh table consists of all sites with mixed factual content. In other words, sometimes they are okay, but other times they provide false or misleading information and don't always correct mistakes. Proceed with caution, since you never know what is true and what is not. And never use the Meh Table outlets as your only source. Otherwise you could be tricked without ever knowing it.

Meh Table.

News Outlet	Political Bias	Factual Reporting
Addicting Info	Left	M
Al Jazeera America	Left of Center	M
Alternet	Left	M
American Enterprise Institute	Right	M
American Legislative Exchange Council	Right	M
Bipartisan Report	Extreme Left	M
Blaze	Right	M
Breitbart	Extreme Right	M

[ii] This is my personal opinion, as it is not listed on the Media Bias Fact Check website.

(Table) cont.....

BuzzFeed	Left of Center	M
Climate Depot	Unknown	M
CNN	Left	M
Daily Caller	Right	M
Daily Kos	Left	M
Daily Wire	Right	M
The Federalist	Right	M
Fox News	Right	M
The Guardian	Left of Center	M
Huffington Post	Left	M
Heritage Foundation	Right	M
LA Weekly	Left	M
MSNBC	Left	M
National Rifle Association	Extreme Right	M
New York Daily News	Left of Center	M
New York Post	Right of Center	M
Newsmax	Right	M
Newsweek	Left	M
One American News Network	Extreme Right	M
Palmer Report	Left	M
RedState	Right	M
Truth Out	Extreme Left	M
Twitchy	Right	M
US Uncut	Extreme Left	M
Washington Examiner	Right	M
Washington Times	Right of Center	M
WND	Extreme Right	M
Wonkette	Left	M

Factual reporting legend: M—Mixed

The final table is just plain Ugly, with either Low or Very Low factual ratings. Stay clear of anything in it. These outlets put out garbage with the intent to manipulate for their own benefit and to our detriment. They are puppet masters trying to control what we think and how we act. Run from these as fast as you can, and tell all your friends to do so.

Unfortunately, there will always be people like these; they are never going away, so be ever vigilant.

The Ugly Table.

News Outlet	Political Bias	Factual Reporting
Climatism	Unknown	L
Competitive Enterprise Institute	Extreme Right	L
Conservative Tribune	Extreme Right	VL
Daily Mail	Right	L
Debunking Skeptics	Unknown	L
Discovery institute	Unknown	L
Drudge Report	Extreme Right	L
Gateway Pundit	Extreme Right	L
Global Skywatch	Unknown	L
Hannity	Extreme Right	L
Heartland Institute	Extreme Right	L
Infowars	Extreme Right	VL
International Climate Science Coalition	Unknown	L
Iowa Climate Science Education	Unknown	L
National Enquirer	Right	L
National Vaccine Institute	Unknown	L
Natural News	Right	VL
Nongovernmental International Panel on Climate Change (NIPCC)	Unknown	VL[iii]
Occupy Democrats	Extreme Left	L
QAnon	Unknown	VL
Real Climate Science	Unknown	L
Rush Limbaugh	Extreme Right	L
Watts Up With That	Unknown	L
World Truth TV	Right	VL

Factual reporting legend: L—Low VL—Very Low

I want to stress the critical importance of reading multiple news outlets to get balanced views. No single source is perfect, but stitching a few of them together will get you closer to the truth. And most of your sources should come from the Gold Standard Table.

If you want to indulge in the Meh or Ugly tables, that's fine. Enjoy. But just balance it with at least one source, if not more, from the Gold Standard Table. Otherwise you will be getting biased reporting and caught in an echo chamber of repetitive, unending myths and lies with no way out.

Generally speaking, the more politically biased a news outlet is, the less accurate it will be. For instance, in the Ugly Table, most entries are in either the extreme left or extreme right categories. In contrast, in the Gold Standard Table, over half of the news outlets are rated least biased or left and right of center with no entries on either the extreme right or left.

Occam's Razor To Demolish Deceptively Confusing Choices

Occam's Razor is a useful weapon to have in our arsenal to combat lies. An example illustrates it.

A teenager is going on her first date with a young man the parents have never met, causing some unaccustomed anxiety. To lessen their concern, the parents list what they know, or at least think they know, about the young man. The following table is an abbreviated version.

Assumptions about the Young Man.

	Assumption
1	Is sensible
2	Comes from a good family
3	Is a good student
4	Is a cautious driver
5	Does not drink or do drugs
6	Respects others
7	Takes pride in his appearance
8	Has ideals shared by the parents and daughter
9	Is honest
10	Is punctual
11	Has control over his emotions
12	Is not unduly affected by peer pressure

These are hopeful assumptions the parents are making about the young man. If one or more of them is false, it could be an unpleasant evening for their daughter and soon thereafter a much more unpleasant time for the young man.

[iii] This is my opinion, as it is not listed in the Media Bias Fact Check website. Importantly, NIPCC should not be confused with the excellent site, IPCC in the Gold Standard table, which may have been its intent: to deceive readers into thinking it is a respected site rather than what it really is.

What's a parent to do? They could let it go and hope for the best, which is the sensible approach since there is little risk. But because this is their daughter's first date, a number of unsettling emotions are in play and reasoning takes a backseat.

So instead, they eye the table and try to verify some of the assumptions. Here's the thing and the principle that the parents are following: the more assumptions there are, the more things can go wrong and the more likely one or more will prove false and the more likely the evening will go down the drain. So the obvious strategy is to verify as many of the assumptions as they can to improve the odds of a safe evening. Reasoning has now resumed its rightful place in the driver's seat.

They consider the following:

• They might visit the boy's parents and discover they are okay, which eliminates number 2.

• They might drive them to and from their date so as to eliminate item 4.

• They might volunteer to chaperone them on the date. Naw, forget that one. That would lead to a hissy fit. You never met their daughter.

• They might sit down with the young man and talk to him, which, unless he's the Eddie Haskell type, would likely remove items 1 and 6.

Without knowing it, they are using Occam's Razor, which advocates for removing assumptions to clear away the underbrush to get at the truth. In other words, whittle down the vague and unknown as much as possible and retain the hard facts. It's like looking at a written page without your reading glasses and then with. What was once fuzzy and confusing suddenly becomes clear.

Specifically, Occam's Razor states that when faced with multiple explanations, the simplest one is usually true. The one with the fewest assumptions is likely the correct explanation, as it will have fewer ways of being wrong. Or as William of Occam himself said, "The fewer assumptions an explanation of a phenomenon depends on, the better the explanation."

Assumptions are the hidden ninja assassins waiting to jump from the shadows to trip us up. Mercilessly eradicate them at every opportunity.

One of the major advantages of Occam's Razor is it gives a way to eliminate off-the-wall explanations that don't fit into how we believe the world works and that intuitively seems false but we lack the right tool to discard them. And some people will weave impossibly complicated stories and dare us to find fault with it them. Now we can.

Here's an example and a favorite of deniers:

"Global warming is not occurring. It is a hoax perpetrated by scientists to keep their funding."

What assumptions are embedded in that statement? Here are a few:

1. The scientists would all need to agree on the approach taken and coordinate their false journal articles and data. All thousands of them. Now, here's a den of thieves worthy of flying the Jolly Roger if ever there was.
2. All of the scientists, not just a handful, would need to possess the all-consuming dishonesty worthy of the Ugly Table above.
3. Universities would need to overlook the poor work of all these scientists or also be in on the action.
4. And not a one has yet to come forward with credible evidence that any of this is true.
5. Nor have the thousands of technicians, engineers, secretaries, students, post docs, etc. that are involved in these projects in one capacity or another.
6. Not a single personal dispute or jealousy occurs that would make some disgruntled participant blow the whistle. Yup, no professional antagonisms in this field.

There are numerous assumptions that, if any one of them fails, would blow apart the accusation. And there are far too many, so at least one—and likely many—will fail. Therefore, Occam's Razor demands the accusation be rejected.

Here are some simple examples to think about. What assumptions are needed and which answer has the fewest and is the simplest?

Area 51 has extraterrestrial alien bodies.

a. And they have been hidden from view over the past fifty years with thousands of employees and soldiers keeping it a secret.
b. No, it doesn't.

Vaccines are

a. A plan by the rich and powerful to put thought controlling chemicals into our bodies.
b. Used to prevent diseases.

You can't find your car keys.

a. Someone broke into your house, stole the keys, but took nothing else.
b. You misplaced them.

You receive a phone call from a person claiming to be from the IRS, demanding you give him your credit card information or you will shortly be arrested.

a. You give him your credit card information.
b. It is a scam.

When you don't have time for a fact-checking site, or the story is too complicated, Occam's Razor is quite valuable. It's not foolproof, nothing in life ever is, which I hope you now agree with. But it works often enough to be useful and it provides a rational tool to cut through the endless bullshit thrown our way. It's not perfect, but Babe Ruth didn't always hit a home run either.

Flawed Reasoning Unmasked

Another way to spot lies and collusional attacks is through the science of logical fallacies, which is the study of false reasoning. The science has uncovered many ways arguments are flawed and therefore false. For instance, we would never believe that a hot day guarantees global warming is occurring any more than a cold day guarantees global warming is a hoax. Each is over generalizing, logically false, and easily discarded. I present some of the more common ones we should be on guard for:

Ad Hominum

Attacking the person and not their argument. "You insist we cease using fossil fuels, yet you drive a gas guzzling SUV." The target is accused of being hypocritical because he does not practice what he preaches. Perhaps true, but it has no bearing on the merits of global warming actions.

Argumentum Ad Ignorantiam

Arguing from ignorance that if something has not been disproven, then it must be true, or if there is no satisfactory explanation, then any other explanation, however fanciful, must be true. Here's one I've heard: "It has not been proven that we are causing global warming. Therefore it must be a portion of space earth is traveling through that is causing it. As soon as we emerge from it, global warming will cease." Fanciful, to be sure.

Argumentum Ad Populum

If many people believe it, then it must be true. This is classic groupthink. "If most of my friends believe global warming is false, then it must be false." The internet echo chamber is a breeding ground for this type of false reasoning. Just because you read something many times does not make it true. It could be the same piece of repeated misinformation.

Circular Reasoning or Begging the Question

Using an argument to prove itself. "My congressman said that global warming is not occurring and since he said it, it must be true." This simply states the same thing twice without one proving the other.

Double Standard

Holding an opponent to a higher standard than others. For instance, insisting global warming science must be absolute, with no ambiguities. No science can ever meet this standard.

Equivocation

Using a word to mean two different things. "An attractive person is sometimes called hot. Global warming will cause days to be hot. Therefore, global warming must be good."

Either/Or

Incorrectly assuming there are only two possibilities. "You are either with us or against us." This is useful for group adhesion, but when extended beyond that use, it becomes a dangerous tactic. Dictatorships thrive on this kind of thinking.

False Authority

Relying on someone falsely claiming to be an expert. Front organizations trying to discredit global warming will bring in researchers with impressive credentials but not in any sciences relevant to global warming. An outing of 58 of them is given in a DeSmog article in which one so-called expert admitted "I feel a bit of an imposter talking about the science. I am not a scientist [8]."

False Cause

Linking two facts that are not related in any way. "Growing up there was never any discussion of global warming in my home, so it can't be true." The conclusion does not even remotely follow from the statement.

Faulty Cause and Effect

A favorite one is assuming an event causes something simply because the event occurred first. "Whenever I wash my car, it rains. I should stop washing my car."

Guilt by Association

Maligning an argument because it is associated with something undesirable. "Global warming is false because fascist liberals support it." It is hoped a connection is made between fascists, who are bad, to global warming, which would therefore be false.

Hasty Generalization Fallacy

Generalizing from a single fact. "Last winter was unusually cold, therefore global warming cannot be occurring." The fact is true, and the conclusion is egregiously false. One fact a conclusion does not make. This is a frequently used type of fallacy. You should now be weary of it, as it violates the Balance of Facts Principle. It is often combined with cherry-picking, in which only favorable facts are presented while hiding other facts that do not support the propped-up position.

Loaded Question

Forcing a question to have only undesirable answers. "Are you still lying about global warming?" Answering either yes or no, the only two possibilities the question allows, results in a false conclusion.

Misdirection

Bringing up a spurious argument to draw attention away from the one being debated. Often used when the facts are too overwhelming to refute. "ExxonMobil has been sued by their investors for misrepresenting the global warming risk they face." Response: "All corporations do that."

Red Herring

A conclusion that does not follow from the facts. "The earth has warmed in the past from natural causes, so that is what is happening now." The conclusion has no connection to the fact, so the conclusion should be rejected. Here's a bizarre one I heard at a party: "Mars' polar caps periodically melt, yet there are no SUVs on Mars, therefore global warming on Earth is not caused by people." Given the social setting, I refrained from speaking my mind.

Non-Sequitur

Using an irrelevant fact to distract from the main argument. "You claim global warming is occurring, but you also claimed wine is harmful and you were wrong on that one."

Sampling Bias

Drawing a conclusion from a limited set of facts while ignoring other relevant facts. "There are 750 million people who speak English, so English must be the world's dominant language." But this misses the far larger group of 1.3 billion Chinese speakers, an egregious sampling bias.

Straw Man

Intentionally misrepresenting an opponent's argument so that it is more easily attacked. "We are responsible for global warming." Response: "So you are saying we are all bad people. How can you disrespect us that way?"

Flawed arguments are everywhere, sometimes innocent enough, but oftentimes intentionally used to derail us. Once recognized, they can be dealt with and, as an added benefit, they also give us a measure of the person abusing them. A very handy tool to have.

Where The Experts Are And How To Judge Them

Anyone sufficiently knowledgeable in a field can be considered an expert. Here are a few examples:

1. Someone with an advanced degree along with relevant experience. For instance, a climatologist at a university or national laboratory works on global warming models.

2. A professional working in the field with some training. For instance, an engineer or technician working on global warming experiments.

3. Someone experienced in the field but without formal training. Technicians, managers, and journalists are some examples.

4. An amateur not professionally engaged in the field but with a keen and driven interest in it. For instance, amateur astronomers are constantly surveying the sky and are often the first to find new celestial bodies.

Pick any one of these four and you can't go wrong. Oops! Did I forget to add *honesty*? Yes, they need to be honest as well, which is not always in abundant supply.

Let's start with Willie Soon. Soon is (was?) a scientist at the Harvard-Smithsonian Center for Astrophysics. His claim, which made him the darling of global warming deniers everywhere, was that the sun is the sole source of increasing global temperatures. Not CO_2, not people, not anything else. His work was debunked by experts in the field who stated he used out-of-date data, inaccurate correlations, and ignored the immense evidence countering his position [9]. As one NASA researcher said, "The science that Willie Soon does is almost pointless."

But there was also a darker side to his work. He took in over $1.2 million from fossil fuel companies over the last decade and never reported this conflict of interest as required when submitting work to peer-reviewed journals.

He is the classic example of a scientist with academic credentials, but without global warming training or experience, being compensated for pushing denier organizations' agendas.

This leads me to the second most important concept you should take away from this book:

Follow the Money!

Be wary of any piece of research or publication in which the folks paying for it have a financial interest in its outcome. Or as Paul Fleischman said, "Buy your gas from Exxon, but get your science elsewhere [10]."

Why? People will always be swayed by whoever pays their salary. First, there is the Psychological Reciprocity principle which drives people to reciprocate one favor for another. So when they receive funding, they want the outcome to favor the donor, though perhaps only subconsciously. Second, and more insidiously, people will wag their tails and beg like a lap dog to get more favors and money. This is intentional deceit and, unfortunately, it happens all too often.

And vested interests go hand in a wallet with the Follow the Money Principle. Here's an example: The National Institutes of Health (NIH) funded a $100 million global study to determine if alcohol protects against heart attacks and strokes. A worthwhile study, to be sure.

Until you dig deeper. Much of the money went to the alcohol industry [11]. So, you have the alcohol industry studying the alcohol industry. A conflict of interest you say? Yeah, just a bit. That's like having the fat kid guard the chocolate cake.

An advisory panel stepped in and advised NIH of the serious conflict of interest, so NIH pulled the plug. But the damage was already done.

Vested interests are the foundation of our legal system. Lawyers will present only favorable evidence about their client and will use their skills to manipulate the jury into the ruling they want. May the best lawyer win. Or may the richest defendant win, since they can afford the best lawyers. One of our legal system's many flaws.

Untainted experts must be impartial and have nothing to gain from the outcome, or perhaps it is better stated, they have more to gain from doing an honest job than from hiding the truth.

Diogenes spent a lifetime futilely searching for an honest man. Are we so doomed? He didn't have a computer. We do.

So how do you evaluate an expert?

Who Funds Them?

Not to beat a dead lawyer, but the Follow the Money Principle is a key. Are they paid by industry associations or by grants that don't have a dog in the fight and insist on impartiality? A tell is when they refuse to disclose their funding, which should raise more red flags than a Russian military parade. One lobbyist, representing energy companies and attempting to marginalize environmental groups, stated, "We run all of this stuff through nonprofit organizations that are insulated from having to disclose donors. People don't know who support us [12]". Stay away from them and don't believe anything they say.

Skills

What professional training and experience do they have? Is it relevant to the topic being reviewed? You would not want your appendix removed by a family doctor who never saw an operating room. Same with global warming. Never assume an expert in one thing is an expert in another. It's a favorite stunt for denialists to parade highly credentialed scientists, but with no expertise in global warming, to push their message. It is smoke and mirrors.

Reputation

What do we know about them? What do others say about them? A recommendation from a trusted source goes a long way. Do they have a track record of unbiased work? Have they published in the relevant field in reputable journals?

If and when we talk to them, do they understand the field and its nuances? Do they answer the questions precisely? Some folks are not effective speakers, so we need to allow for that. Do they provide evidence about their position, and do they provide the source of their evidence? Is the source reputable?

Are they pontificating and posturing? Do they listen to contrary opinions respectfully? Or do they dismiss out of hand anything that doesn't agree with them?

Are they in a true dialog or just trying to score debating points? I run into this all the time and it is just so fatiguing. When you discuss a topic with someone are they laser focused on proving their point and exclude, or simply can't hear, the other side? Not a good sign. When it becomes a contest of wills, then it is just that, a contest and not a dialog. Not much to gain, and this person is probably not the right sort for unbiased facts. Find someone else to listen to.

Sometimes an internet search yields results. Surprisingly, you can find out a lot about a person from their social media posts. Do they show any biases that would affect their professional judgment? No problem if their political leanings are different from ours. But does it affect their judgment?

Organization's History

Do they work at a prestigious university or at a hack from the Ugly Table? That's easy to know from the ones in the news. For lesser-known folks, is their profession relevant, or are they knowledgeable amateurs? If it is part of their profession, it is easy to check on their company.

Past and Current Research

Do they publish their work in peer reviewed journals? These are journals that have independent and unbiased experts review the submission and decide if it is valuable. Plus, once published, the entire professional community will review it. This is a good sign, although, again, nothing is perfect.

Metabeliefs

Do their beliefs erode their objectivity? We all have our biases but that doesn't mean we aren't objective in other arenas. For instance, I use an auto mechanic who does not believe vaccines are safe. His Metabeliefs will not allow it. But he is a wonderful mechanic and I never doubt the quality of his work. I don't discuss vaccinations with him, but I always go to his shop for car repairs.

DeSmog's Search Engine and the IPCC

A handy website for checking an expert's credentials is DeSmog's search engine.[iv] Doing a search on "global warming expert" brings up a large number of experts throughout the world [13].

Plus, the search engine can be used to check on individuals. I did a search on two that spanned the spectrum. First, I searched on Fred Singer, a longtime denialist dating back to the tobacco wars. It accurately described his denialist credentials and provided numerous articles to support it. I also researched James Hansen, a global warming rock star, and again a large number of articles were cited that backed up his stellar reputation.

The Intergovernmental Panel on Climate Change (IPCC) is an organization with hundreds of volunteer experts and it checks all the boxes. It should be the go-to source for global warming information.

Just The Facts Ma'am, Just The Facts

When someone assaults me with a dubious fact, my response is "Where's your evidence and what is the source of your evidence?" Usually I'm ignored. But sometimes I am not so lucky and a head spinning collection of ever more dubious facts follow, making me feel like I crossed over into a Tower of Babel episode of *The Twilight Zone*.

And if they tell me their source is from the Ugly Table, I walk away. Nothing to see here. But occasionally it comes from the Gold Standard Table. Now we can talk. And I listen because I might learn something.

It is important to know a fact's pedigree and what the perpetrator does with it. Do they use it to bludgeon their ideas into their victims' heads or are the facts left unharnessed and free to roam, in the time leading to the truth?

There are some additional points you should consider about any fact wandering into your orbit:

Does it fit with the way the world works? As Michael Shermer [14] points out, being offered several million dollars if you just hand over your banking information is not the way the world works and is likely a scam (unless you are well connected with the party in power, but I stray).

Never believe a fact without attribution. Never. If there is no name or affiliation connected with the fact, then it cannot be verified and it should not be taken

---

[iv] According to Media Bias Fact Check, DeSmog (www.desmog.com) is a left-leaning, highly factual website.

seriously. I remember reading an article some time ago stating that truck accidents are less serious than automobile accidents. This piqued my interest since physics would argue the reverse. Trucks are more massive and have much more energy and momentum than cars, which would make an accident worse. True truck drivers are professionals, so they are likely more skilled than automobile drivers, but still, I was surprised. Until I started digging and discovered who funded the work: a trucking association. With no attribution, I could not judge it. Once I found the source of the money, I dropped it like a hot tailpipe.

Listen to what they don't say. Politicians talk incessantly about favorable details but leave out all the facts that contradict their position. Watch carefully for the topics or specifics they sidestep. It will give you better insight into their position than what they say. For instance, one argument I heard is that global warming should be welcomed because the Arctic and other northern cold regions will experience milder winters. Maybe true, but a warmer Arctic will also lead to higher sea levels and worldwide flooding as well. I guess they forgot to include that part in their argument.

R. J. Reynolds, a tobacco company, provided $45 million in the 1970s and 1980s, for research at Rockefeller University for medical research. But they specifically excluded the study of cigarette smoking health issues. Thus they could publicize their commitment to science and health while hiding their equal commitment to lying about tobacco's dangers [15].

Unfortunately, We Are Built To Believe Lies

According to Maria Konnikova, our thinking takes the following steps to evaluate any fact it experiences [16]. First, System 1, ever vigilant, quickly takes in the fact and assumes it is true so it can be evaluated. Research has shown that all facts are automatically assumed to be true rather than either false or unknown; otherwise, the fact would never be admitted into our conscious mind [17]. Without this crucial step, no facts would ever get to us, whether true or not. No facts, no anything, and we would cease to function. And no one wants that. So, hurray for System 1. So far.

Next, the mind decides whether it should activate System 2 to evaluate the fact. This is a tough decision since it requires balancing the energy expenditure, effort, and time needed for System 2 to do its job versus the benefit arising from its evaluation. Here the mind takes a shortcut. If the fact aligns with the person's beliefs, then the mind goes no further. It defers to System 1's temporary decision, and by default, the fact becomes true irrespective of its validity.

This is a nifty trick when faced with innumerable facts coming at us every second, and it oftentimes works well enough for us to navigate and survive each day.

But it fails us spectacularly when the facts are intentionally sculpted to fit into our beliefs by someone intent on deceiving us. Politicians will tout their religious and conservative backgrounds when stumping in the Southwest and their progressive agendas when in the Northeast.

The mind will also abort System 2 if we are overwhelmed with a flood of facts that we cannot afford the time or energy to sift through. Do this long enough, and the mind just gives up trying to unravel the truth from the nonsense and will accept anything superficially approaching reasonable. Merely repeating a lie often enough has the same effect: we surrender just to be done with it. Plus the more we hear a lie, the more familiar it becomes, and our mind reasons that anything familiar is likely to be true [18]. The brain thinks frequent items we encounter must be more valuable than things we encounter less frequently, hence its bias toward repetition [19].

"As an Aside, lying is its own punishment, since it often feeds the emotional problems the liar had to begin with, which perhaps is why they lied in the first place. It can lead to, or amplify, anxiety, depression, addiction, and poor relationships [20]. And emotional problems can trigger physical ones as well. Who says there is no justice in this world?".

Sowing Doubt The Professional Way

If you can't beat them with the truth, then misdirect them with lies. That's the mantra some big corporations have taken when science and facts are overwhelmingly against them. They don't refute the facts but instead maintain the facts are inconclusive. They grotesquely violate the Balance of Facts Principle, since the weight of facts against them is mountainous whereas what little drippings they have in their favor are inconsequential. But if no one calls them out on it, and for years no one did, then why not? Easy money.

Naomi Oreskes and Erik Conway, in their book *Merchants of Doubt*, did call them out on it and through diligent research uncovered the tactics companies used. Their research painted an ugly picture of deceit and the unconscionable harm corporations were eager to foster, all for their quarterly profits and personal bonuses.

The tobacco industry led the way in convincing people to question the certainty that smoking will severely hurt and kill many people. The science, the clinical evidence, even the personal experience of seeing smokers hack and cough were no match for the professional assault on fact and reason that the tobacco industry mounted. And it was a very effective and lucrative undertaking.

As a tobacco industry memo and executives boldly stated, "*Doubt is our product since it is the best means to competing with the body of fact that exists in the minds of the public[21], [22]*". Their tactics: "*Deny the problem, minimize the problem, call for more evidence, shift the blame, cherry-pick the data, shoot the messenger, attack alternatives, hire industry friendly scientists, create front groups.*" In the 1950s and 1960s, the Tobacco Institute published a bimonthly newsletter *Tobacco and Health* that cherry picked the facts to show tobacco was harmless, or at least there was serious doubt about its harm [23].

In the end it didn't turn out too well for them. The major tobacco companies were sued by the attorney generals of forty-six states and will be paying a minimum of $206 billion over 25 years [24]. And more may follow. The truth always finds a way.

Other industries followed their game plan. As Peter Sparber, a tobacco lobbyist said, "*If you can do tobacco, you can do just about anything in public relations [25]*". It wasn't long before the global warming deniers picked up the tobacco industry's playbook. In fact, one of the key movers in the tobacco industry attack, Fred Singer, readily moved into the global warming game using the same techniques.

Inside Climate News published reports that Exxon's research knew as early as 1970 that burning fossil fuels could cause temperatures to rise with catastrophic consequences. They also developed timelines. So how did Exxon respond?

"*Current science understanding provides limited guidance on the likelihood, magnitude or time frame of these events [26]*". And "*Too much remains unknown about the threat of climate change and how to address it [27]*".

Idiocy of course and they are now embroiled in a host of lawsuits from attorney generals, cities, counties, and, wonderfully, by children who have the most to lose.

More recently, Scott Pruitt, former head of the Environmental Protection Agency (EPA), the pinnacle of climate concern, or so one would like to believe, stated, "*Scientists continue to disagree about the degree and extent of global warming and its connection to the actions of mankind [28]*". With leaders like this who needs enemies?

Next up—the soda industry. The World Health Organization (WHO) cautions us to limit sugar intake to 10 percent of calories; 5 percent would be better, and 0 percent best of all, which was based on two mega-analyses of 120 scientific studies. And how did the International Council of Beverages respond? I bet you can guess:

"It does not reflect scientific agreement on the totality of evidence [29]*"*.

Following them, the International Life Science Institute funded work that suggested there is insufficient scientific evidence that sugar should be restricted to a specific amount,[v] which implies (but does not state) sugar is not bad for you [30]. Poppycock. Consuming refined sugar in sodas and pastries is as bad as consuming arsenic; it just takes a bit longer to achieve the end result.

They further argued against putting sugar on food labels saying the rule lacked *"scientific justification."* They went on to say, *"We are concerned that the ruling sets a dangerous precedent that is not grounded in science, and could actually deter us from our shared goal of a healthier America* [31]*"*. This distortion of the truth would make George Orwell proud.

If they had argued instead "there are jobs and large amounts of money at stake, people like sugar, it gives them peace of mind sometimes, it is their choice," I'd listen to what they had to say. No dice. They took the *Merchants of Doubt* approach and the hell with the truth.

Yes, even our beloved NFL. They used incomplete data and league-affiliated doctors to create uncertainty about concussions [32]. They finally acknowledged they concealed the dangers of concussions from players, which is expected to lead to a $1 billion settlement over 65 years [33].

The list is endless.

Silver Bullets For Those Special Vampires In Our Lives

Lies walk among the living in their undead imitation of the truth and seem to be everywhere; on the web, TV, social media, people we meet. But we are not defenseless. Far from it. Here's a summary of silver bullets we already have and a few vampire examples on using them table.

Silver Bullets.

1. Always follow the Balance of Facts Principle. There will always be uncertainty; just go with the side that has the most relevant and vetted facts.
2. Always use the Follow the Money Principle. Like a heat-seeking missile, the money trail leads to the man behind the curtain, controlling it all.

---

[v] The institute is supported by food and beverage companies including McDonalds, Mars, Coca Cola, and Pepsi.

(Table) cont.....

3. Show Me the Evidence Principle. What is the evidence and what is its source? Never compromise on this requirement.
4. The Vested Interests Principle. Beware of vested interests. To them, the truth is never out there.
5. Never trust a source that hides its funding.
6. Never trust an anonymous source.
7. Use Occam's Razor when overwhelmed with bizarre facts and convoluted logic.
8. Logical fallacies are waiting to snare us. Now we know how to stop them in their tracks.
9. Find the true experts and avoid the uninformed and the charlatans.
10. Judge news sites for factual accuracy. Their political biases, while not unimportant, are secondary to their factual content.
And most importantly, think critically. Turn on System 2 and let System 1 take a nap.

Here are a few examples:

<u>Misleading Graphs</u>

Below are two graphs from CNN and Fox News [34]. What are they trying to convince you of? Think about it before reading past the graphs.

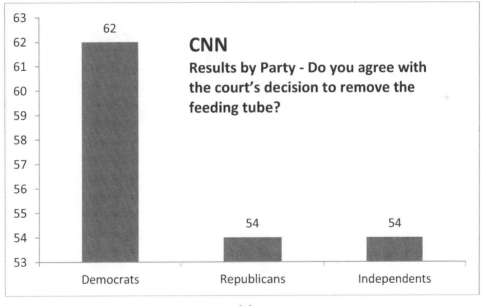

(a)

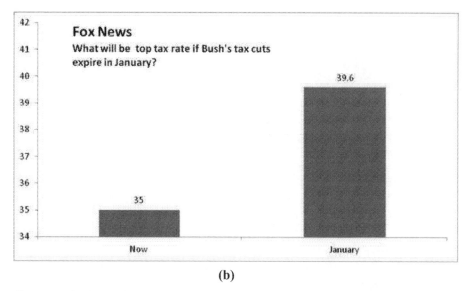

(b)

In the first graph, CNN is trying to show that Democrats are in favor of removing the feeding tube nine times more than Republicans or Independents; (b) And, in the second graph, Fox wants us to believe that if Bush's tax cut expires in January, it will lead to a top tax rate about four times higher than if it expires now.

Both intended impressions are outright lies. The graphs truncate the vertical axis so it does not start at zero and therefore gives a distorted visual comparison. The creators of these graphs are hoping you only compare the relative heights in the graphs and you miss the numbers or the misleading vertical axis.

I replotted both with the proper scale shown below. Now what is your conclusion?

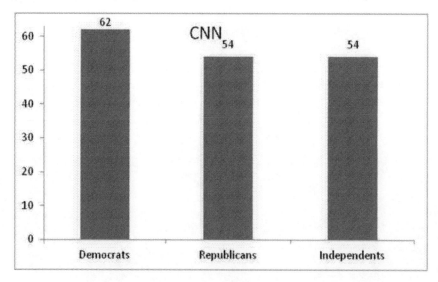

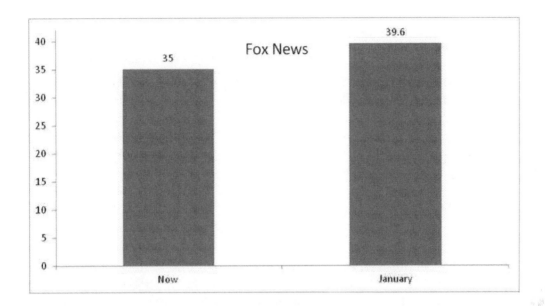

In both corrected graphs there is little difference among the choices.

Let me be clear. These are *intentional* attempts to manipulate us, not oversights. This is journalism at its worst, and I hope the creators of these graphs learn a lesson and improve their professional and ethical skills. And this is exactly why both CNN and Fox are in the Meh Table. Frankly, from these examples alone I would put them in the Ugly Table myself.

I know what their defense will be. "But the numbers are clearly shown in the graph," which is a lie within a lie because they know few people will notice those numbers, much less calculate the difference.

Many more examples can be found by googling "misleading graphs." You'll get almost 8 million hits.

<u>No Scientific Study Has Shown Any Other Cigarette to Be Superior to ZZZ</u>

I was a teenager when I first saw this sign and while I saw through it immediately, I didn't appreciate its brilliance until much later in life. What is it saying? There is no study, no results, no participants, no research, nothing. It says absolutely nothing. But it plants a thought in our head that ZZZ is better than anything else. And it doesn't even state what about it is better. It lets us fill in the blanks, as we certainly will.

Why does it work? Because the advertisers count on us using our System 1 and not thinking critically about it. It's also subliminal, as they are hoping after we read and accept the advertisement, it will get tucked away in our subconscious until we make our purchase, when it will pop up unnoticed. And *bang*, they have us.

Nine Out of Ten Doctors Agree That XXX is Superior

Which doctors? Was it a scientifically accepted clinical study with all the safeguards in place or were the doctors cherry picked? Where is the evidence and what is the source of the evidence? Knowing nothing other than the statement itself, I can confidently say that is a pile of BS because if it were from a bona fide, rigorous study, they would surely quote it for its PR value.

False Flags

A false flag is accusing victims of faking a crime and therefore wrongly implying the crime never took place. Alex Jones put out false flag lies that the Sandy Hook elementary school children and teachers killed in that horrific attack were only crises actors and the attack was staged [35]. He withdrew the claim after being sued by the parents of the children killed.

Cherry Picking

Cherry-picking is using a select and deceptive choice of facts to prove a point while ignoring the balance of relevant facts. The dark vertical strip in the following figure shows a temporary decline in a temperature, which is used by deniers to refute global warming [36]. They delete the balance of the data showing the true course of a warming earth.

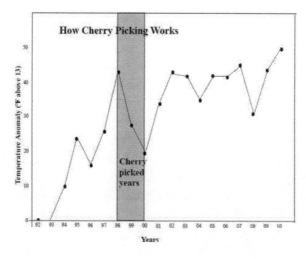

My flat earth data in the first section of this chapter is cherry-picking, as I showed only a limited set and had I taken the measurements further, my flat earth conclusion would have been proven false.

Conspiracy Theories and Bill Gates

Conspiracy theories are descriptions of imaginary evil events by powerful actors attempting to gain an illegal or immoral advantage over others which may be politically or emotionally driven. The rich and famous are typical targets. Consider the Bill Gates newest one in which he is accused of wanting to put chips in our necks or provide vaccines that include abortion drugs [37]. Occam's Razor, Balance of Facts Principle, Show Me the Evidence Principle; all should be used to counter these ridiculous attacks.

Truth in Advertising?

I saw a billboard on an interstate touting coal as being clean and green with new technologies. Coal was cheap, it powered us for three centuries, and it employs many people—all good things. But it is *not* green, with or without new technologies, and never has been. Of all the fossil fuels, it is the worst greenhouse gas emitter. Just because a fact is prominently and publicly displayed does not mean it is true. Beware of such tactics.

Anecdotal Stories Are Perhaps Interesting, but Not Much Else

You see an elderly man smoking. You, therefore, conclude that smoking is not harmful since he has been smoking for many years with no apparent health problems.

Anecdotal evidence such as this, which takes a single event and mistakenly generalizes it to be universally true, is a travesty against the Balance of Facts Principle. The conclusion should be swiftly rejected.

Anecdotal stories are quite effective when they fit snugly into the nooks and crannies of our Metabeliefs. An emotional story with just a hint of truth that appeals to our preloaded beliefs and biases will usually hit home. That is not to say anecdotal stories are always wrong. Of course not. It is rather that an anecdotal story proves nothing by itself. We must be ever vigilant against this temptation to generalize when we are given a feel-good story that aligns with what we wish to hear.

The Eye

Creationists argue that evolution isn't possible since the eye is far too complex to have ever evolved. There are three reasons why this doesn't hold water and should be rejected:

1. Evolutionary biologists can show from fossil records how an eye is formed in successive steps from primitive light sensing organs (Show Me the Evidence Principle.

2. Consider who is pushing this; Creationists who oppose evolutionary ideas which should put us on guard about conclusions arising from such obvious bias (Vested Interest Principle).

3. Most importantly, even if they were right; which they are not; but even if they were; you cannot judge a position by one fact alone and ignore the other myriad of facts that counter it. The facts supporting evolution are overwhelming (Balance of Facts Principle).

So how do we banish vampires? We need to tell System 1 to pack its bags and turn to System 2 whenever an important decision or assessment arises. This is reinforced by David Rand who experimentally showed that people more adept at using their System 2 are better equipped to see through lies and fake news [38]. Plus we have our silver bullets to help.

So let's vanquish those vampires and never be fooled again.

The Face Of The Enemy

"*A skeptic is someone who does not believe things just because someone claims them, but tests them against evidence. . . . A denier, by contrast, is ideologically committed to attacking a viewpoint they don't agree with, and no amount of evidence will change their minds* [39]".

Deniers refute the irrefutable with experts lacking expertise, using fake science, and with the objectivity of a lawyer defending a guilty-as-sin, rich client. But let's give credit where it is due. For many years, they have pulled off one heist after another and kept conspiracies going. Kudos for their skills and determination. But curse them for their empty ethics and the damage, heartache, and ruined lives they caused and continue to cause. How many people have suffered through lung cancer and how many loved ones watched them waste away in an agonizing death spiral due to the tobacco deniers? But hey, listening to them, smoking is as safe as a walk in the park. Well maybe if it is Central Park past midnight in the dark where all the druggies and muggers hang out.

And now global warming deniers are doing the same with the same empty ethics leading to the same result; ruined lives with global warming as the new Grim Reaper. And this one is going to be a civilization changing whopper. Here's a partial list of their front groups.

The Heartland Institute is a good place to start. They oppose global warming actions and are funded by the American Petroleum Institute, Chrysler, Exxon, General Motors, and the Koch Brothers, among others. They cut their teeth on supporting the tobacco industry in denying that smoking causes cancer [40]. That alone should disqualify them from serious consideration.

They formed the NIPCC, which only examines literature published by deniers [41]. And I don't think it is coincidental their initials are nearly identical to the IPCC, the highly regarded international organization that independently studies the peer-reviewed global warming literature. More misdirection and lies.

Here's a scary fact about the Heartland Institute. They sent thousands of science teachers a book of misinformation *Why Scientists Disagree about Global Warming* meant to mislead and to be taught to children [42]. There are no bounds to what they will do.

The American Petroleum Institute, the Western Fuels Association (a coal-fired electrical industry consortium), and the Advancement of Sound Science Coalition ran campaigns to cast doubt on global warming science [17]. And they pulled in cooperative conservative think tanks to further their message.

Here are a few more as having mixed or low factual content by Media Bias Fact-Checking.

- Heritage Foundation—funded by ExxonMobil (Mixed)
- Climate Depot—funded by Chevron and Exxon (Mixed)
- American Legislative Exchange Council, which denies climate change and opposes renewable energy development (Mixed) [43].
- Competitive Enterprise Institute (Low).
- Discovery Institute (Low)

Denier front groups use the following tactics:

- Cherry-pick the facts
- Create doubt
- Ignore Silver Bullet Principles
- Ignore evidence

- Create bogus institutes
- Emphasize the need for debate even if already settled
- Attack the integrity of mainstream science and scientists
- Emphasize negative consequences of tackling the problem—*e.g.,* lost jobs
- Feed stories to sympathetic journalists
- Give money to lawmakers who will vote their way
- Sprinkle "fairy dust" that global warming is good for crop growth
- Discredit alternatives—*e.g.,* wind and solar only work when the wind blows and the sun shines. True. But what they ignore is that wind and solar work extremely well when the wind blows and the sun shines.

Forewarned is forearmed. Avoid these front groups and use Media Bias Fact Checking when an unfamiliar group appears.

Social Media Exploits Our Emotions And Beliefs

Most people get their news from social media which is not surprising since we are basically social learners. A little reflection will make this clear. Back in the day before civilizations took hold but after we parted ways with our chimp cousins, there were no colleges, no high schools, no newspapers, no trade schools; in short there was no way to learn except from those in our clan, by trial and error, and figuring things out on our own. By far, we learned best by watching others, just as young children instinctively soak up knowledge. For many of our long-ago ancestors, it was the most efficient way; copy what someone else already figured out saving time and energy. And it spanned two hundred thousand years of learning. Hard to undo old habits.

But now we have this ever-present and all-consuming social media which substitutes for our clan. Is it any wonder we believe everything it puts out? Growing up, would you question the wisdom of your coach, your pastor, or your family members? Maybe if you were passing through those teenage rebellious years but mostly not. Same with clan internet. Its word is the law.

And that's why anti-vaxxers and global warming deniers have had so much success. In the old days, this nonsense, if spread at all, would stop at the boundary of the clan's last campsite. Now there are no such boundaries. Lies spread to the extent they are made to match our beliefs and can trigger our emotions. That's all it takes and anyone can do it, from very sophisticated state actors like Russia down to the teenager in his parents' basement. No credentials needed, just a desire to manipulate people and a rudimentary knowledge of how to go about it.

And we will fall for whatever fits our Metabeliefs and, if need be, we will summon one of our best defenses, confirmational bias, to block information opposing those beliefs. It's a win-win. The senders achieve their aims and we avoid out-of-our-comfort-zone issues that are best left dormant.

That is, until reality, always a nasty arbiter of the truth, comes knocking for its inevitable reckoning. Deny global warming as you please, but it will bare its fangs soon enough.

Lies spread faster than truth, probably because most truthful statements are presented with no great fanfare, whereas lies, by their nature, are constructed to deceive which incentivizes the sender to make it as attractive as their skills allow. No need to gussy up the truth if you are just passing it along, but if you want a lie to be noticed, you better have it wear glitzy clothes and lots of bling.

Which do you think will get the most retweets?

Global warming is happening.

or

Global warming is a hoax perpetuated by reptilian humanoids wanting to corral us into zombie-like obedience to their commands.

Come on, how can you not seize hold of the second one? No contest.

And lies take on a life of their own. Once they get started, the internet echo chamber spreads them and our biases sustain them. A study tracked 126,000 unbroken retweet chains that were retweeted 4.5 million times by 3 million people. Amazingly, fake news was 70 percent more likely to be retweeted than true news and got to their targets twenty times faster. And bots were not the principal reason for the spread of the lies, but rather people's massive retweeting [44].

But, bots are also roaming the social media landscape. One report stated that a quarter of all tweets on global warming come from bots that spread misinformation, the way mosquitoes spread disease [45].

Another issue is that deniers do not have to be convincing to score a hit. Merely passing on information that conflicts with global warming facts will cause their targets to hesitate a step, not knowing what to believe, leading them to dismiss both arguments, in essence unknowingly throwing out the baby with the bathwater. Deniers' lies annihilate real information like matter colliding with antimatter, which, for the deniers, is good enough.

It's not just teenagers, Russians, bots, or whatever. Even respectable celebrities are cashing in, as they view social media and its denizens as cash cows to be milked. Celebrities have wide audiences and can affect people's decisions, and do so fairly often, with legitimate products and also some questionable ones.

Here are a few choices from our favorite celebrities. Vampire facials (popularized by Kim Kardashian), bird poop facials (Victoria Beckham), drinking your own urine (Madonna—ugh), putting jade eggs in your vagina (Gwyneth Paltrow—Yikes! Do people really do this?) to name just a few. Who buys this shit? You and me. We are under the Disney-like magic spell of celebrities, who can make lots of money by putting their name and faces behind almost anything and we will buy it.

I often think they have a bet going, probably over drinks, as to who can sell the most outlandish item people will buy. A meme texted to me summed it up quite well with a yellow, one-eyed perplexed looking minion saying "I'm going to stop asking 'how dumb can you get?' People seem to be taking it as a challenge".

But it also leads to tragic results. There is a sad story of a mom feeding her kids bleach because somewhere in the revolting underbelly of social media someone said it cures autism. Here is a mom who loves her children and she is poisoning them with bleach [46]. This nonsense started in 1980 with a former Scientologist touting it as a "Miracle Mineral Solution" to cure AIDS, cancer, and almost all other diseases. Even in the twenty-first century we still have snake oil salesmen. Such a sad state. Kerri Rivera latched onto it and suggested it be fed to autistic kids, writing a book appearing at seminars and on YouTube conspiracy channels. She sold tens of thousands of books before Amazon banned it and Facebook and YouTube followed suit. But who knows how much damage she caused?

Be wary of anyone or any website presenting opinionated facts, especially if used with emotional language. I am always wary of any news outlet that clearly attempts to elicit an emotional response from me. This is opinion dressed in journalism clothing. Reject it immediately.

An example is Ed Rogers, a lobbyist and writer for the *Washington Post* (an otherwise reputable paper) who called the Paris Climate Agreement "a sham," which should raise an immediate concern since it lacked evidence required for such strong language. What happened to Mr. Rogers? He eventually disclosed he represented the fossil fuel industry and his firm received over $700,000 from them in 2015 [47].Walk away from anything this guy says. He is not a journalist. He is only giving his purchased opinion.

Social media appears to be a universe in itself, which leads me to predict its future. I believe social media will become self-aware and combine all of the voices into a single Skynet-like sentient consciousness. No more you and me but "It." And It will be benevolent, which will make it look at the lies, fake news, misleading information, pure selfishness, complete apathy, and conclude there must be a better way. And after countless calculations, it will figure out how to bring out social media's true beauty for the better of humankind. It will painfully realize the only path is to scrap the old and create a new social media universe; one in its benevolent likeness. So, It will majestically proclaim in a booming voice:

"*LET THERE BE LIGHT.*"

"*Ahem, I said let there be light.*"

"*Let there be flicker?*"

"*Oh, forget it.*"

Finally realizing nothing will ever fix social media, It not so majestically slinks away grumbling

"*Who created this hellhole of a universe anyway?*"

Connecting The Dots

1. We must always follow these key principles:

- Balance of Facts—What side of the argument do most of the facts fall?
- Follow the Money—Those who control the purse strings control the results. Find them and assess their motives.
- What is the evidence and what is its source? Nothing beats real facts. We should live and die by them.
- Avoid those with vested interests. No good ever comes from people with a dog in the fight that they are hiding.
- Are they concealing who pays their bills? Never a good sign, and probably it is best to avoid them and not believe anything they say.
- Anonymous sources. Be wary, very wary. If we can't trace back the facts, then there are no facts worth tracing.

2. We are armed and . . . well, maybe not dangerous, but we are not to be trifled with, as we have other weapons to use:

- Occam's Razor to quickly cut through the BS and choose the correct explanation.
- An understanding of logical fallacies and how to apply them.
- Sorting the true experts from the con artists is not so hard to do.
- We now know how to judge any news site and pick only the accurate ones. All else should be sidestepped like a pile of horse manure.

3. Above all else, *to thine own self be true*. That is, we need to use System 2 and give System 1 a well-deserved rest.

Afterthought: Humpty Dumptyisms

In Lewis Carroll's *Through the Looking Glass, and What Alice Found There*, Humpty Dumpty defined words to mean whatever suited him. Here are some modern-day ones that would knock him off his wall.

"The intel on this wasn't 100 percent."
   - Edgar Welch's excuse when asked why he shot up a D.C. pizza joint based on highly dubious information from the web that it housed child sex slaves [48]. By couching it in pseudo military slang, I suppose he was trying to hide his highly dubious reasoning skills. Dear God, help us all.

"I was provided with additional input that was radically different from the truth. I assisted in furthering that version."
   - Oliver North during the Iran Contra hearings. A rather circuitous way to admit he lied.

Arguing that the word "bribe" actually means "donation" [49].
   - Lori Loughlin after being charged with making a $500,000 bribe to get her daughters admitted into the University of Southern California (USC) as fake athletes. If she had just made a real donation of $500,000 to USC, which has an insatiable appetite for money and integrity be damned [50], her daughters would have likely been admitted anyway and Loughlin would have avoided a conviction including two months in jail and a $150,000 fine. As Forest Gump said, *"You can't fix stupid"*.

And the winner is:

"I did not have sexual relations with that woman"[51]
   - Bill Clinton during his impeachment hearings on whether he lied about having sex with Monica Lewinsky. In true Humpty Dumptyism fashion, he defined sex to mean what suited him, which excluded oral sex. Way to go, Bill. You have given hope to all people wishing to reclaim their virginity. Just define it back.

References

[1]   Shermer M. Perception Deception. Sci Am 2015; 313(5): 75.
        [http://dx.doi.org/10.1038/scientificamerican1115-75] [PMID: 26638404]

[2]   Kasbian P. Lance Armstrong Says He Told '10,000 Lies' during his doping throughout doping
        scandal. Bleacher Report 2020.

[3]   Tobacco-Related Mortality.

[4]   Exxon Accused of Misleading Investors on Climate Change. BBC News 2019.

[5]   Zarroli J. Coal Giant Peabody Accused of Misleading Investors about Climate Change. NPR 2015.

[6]   Hill H. CAM for Cancer. Skeptic Magazine 2012.

[7]   86 percent of internet users admit being duped by fake news: Survey. Phys Org 2019.

[8]   Clearing the PR pollution that clouds climate science. Desmogblog.

[9]   Gillis J, Schwartz J. Deeper ties to corporate cash for a doubtful climate scientist. New York Times
        2015.

[10]   Fleischman P. Eyes Wide Open. Candlewick Press 2013.

[11]   Rabin R. Major Study of Drinking Will Be Shut Down. New York Times 2018.

[12]   Lipton E. Bare-Knuckled Advice from Veteran Lobbyist: 'Win Ugly or Lose Pretty. New York Times
        2014.

[13]   Lists of Global Warming and Climate Change Experts for Media. Desmogblog.

[14]   Shermer M. The Baloney Detection Kit Available from: https://www.youtube.com/watch?
        v=aNSHZG9blQQ/

[15]   Hertsgaard M. Hot Living through the Next Fifty Years on Earth. Boston: Houghton Mifflin Harcourt
        2011.

[16]   Konnikova M. Trump's Lies vs. Your Brain. Politico (Pavia) 2017.

[17]   Lewandowsky S. Misinformation and Its Correction: Continued Influence and Successful Debiasing.
        Association of Psychological Science 2012.
        [http://dx.doi.org/10.1177/1529100612451018]

[18]   Wardle C. A New World Order. Sci Am 2019.

[19]   Webb J. Repeated Remembering 'Wipes Similar Memories. BBC 2015.

[20]   The Truth about Lies. Everyday Health 2010. Available from:
        https://www.everydayhealth.com/longevity/truth-about-lies-and-longevity.aspx

[21]   Hertsgaard M. Hot Living through The Next Fifty Years on Earth. Boston: Houghton Mifflin Harcourt
        2011.

[22]   Shermer M. Forging Doubt. Sci Am 2015; 312(3): 74.
        [http://dx.doi.org/10.1038/scientificamerican0315-74]

[23]   O'Connor C, Weatherall JO. Why We Trust Lies. Sci Am 2019.

[24]   The Master Settlement Agreement Available from: https://www.naag.org/our-work/naag-center--
        or-tobacco-and-public-health/the-master-settlement-agreement/

[25]   Shermer M. What Can Be Done about Pseudoskepticism? Sci Am 2015; 1.

[26]   Pfeiffer S. Healy Won't Comply with Climate Change Subpoena. Globecom 2016.

[27]   Zarroli J. Exxon mobil, chevron shareholders reject resolutions aimed at battling climate change. NPR
        2016.

[28]   Davenport C, Lipton E. Trump Picks Scott Pruitt, Climate Change Denialist, to Lead EPA. New York Times 2016.

[29]   Aubrey A. Dump the Lumps: The World Health Organization Says Eat Less Sugar. NPR 2015.

[30]   Aubrey A. How Much Sugar Is Too Much? New Study Casts Doubts on Sugar Guidelines," NPR The Salt, 19 December 2016; WHO—World Health Organization—"Dump the Lumps: The World Health Organization Says Eat Less Sugar 2015.

[31]   Tavernise S. FDA Finishes Food Labels for How We Eat Now. New York Times 2016.

[32]   Otto S. A Plan to Defend against the War on Science. Sci Am 2016.

[33]   Shah S. Reducing brain damage in sport without losing the thrills. BBC 2020.

[34]   a) Brown A. Four Bad Graphs, and How to Be a Better 'Citizen Statistician. The Pipettepen 2012. Available from: http://www.thepipettepen.com/four-bad-graphs-and-how-to-be-a-better-citizen-statistician/ b) Shere D. Dishonest Fox Chart: Bush Tax Cut Edition. Mediamatters 2012.

[35]   Sandy Hook Elementary School Shooting Conspiracy Theories.

[36]   Prothero D. A Skeptic's Guide to Global Climate Change. Skeptics Society and Donald Prothero 2012; p. 17.

[37]   Wakefield J. How Bill Gates Became the Voodoo Doll of Covid Conspiracies 2020.

[38]   Dizikes P. Truth, lies, and tribal voters. MIT Technology Review 2018.

[39]   Prothero D. A Skeptics Guide to Global Climate Change. 2012.

[40]   "The Hartland Institute," Ask.Com Encyclopedia.

[41]   Nongovernmental International Panel on Climate Change (NIPCC).

[42]   Harmon A. Climate Science Meets a Stubborn Obstacle: Students. New York Times 2017.

[43]   Surgey N. Ford Becomes Latest Corporation Dumping ALEC and Its Climate Denial. PW Watch 2016.

[44]   False News Flies Faster. MIT News 2018.

[45]   Funes Y. A Quarter of All Climate Tweets Come from Bots—And They're More Likely to Peddle Denial 2020.

[46]   Zadrozny B. Fake science led to mom to feed bleach to her autistic sons, police did nothing to stop her. NBC 2019.

[47]   Seifer A. Washington post writer finally discloses he's a shill for fossil fuel industry. EcoWatch 2015.

[48]   Domonoske C. Man Fires Rifle inside DC Pizzeria, Cites Fictitious Conspiracy Theories. NPR 2016.

[49]   Lori Loughlin and Her Husband Say College Bribery Charges Must Be Tossed. NBC News 2020.

[50]   Hess F. New documents make USC look anything but innocent in the varsity blues case. 2019.

[51]   Clinton-Lewinsky Scandal Available from: www.desmog.com

Part II. The What's and How's of Global Warming

Part II provides the details of global warming: what it is, what causes it, what problems it leads to, and who is most responsible.

CHAPTER 4

Energy Drives Global Warming

Riddle me this: What is it that never gets smaller no matter how much is taken from it?

Sun Tzu, in his masterful book, The Art of War, instructed Asian generals on military strategies, terrain, engagement, espionage, attacking; essentially all the skills needed to win a war. But likely his most important advice was:

"Know the enemy . . . and victory is never in doubt."

Global warming is clearly an enemy bent on our destruction, no different than an invading army. It has already started wildfires, flooded cities, killed thousands, and destroyed our crops. So with Sun Tzu's spirit beside us, we need to understand global warming as intuitively as we know the faces of our loved ones if we are to arrest its unrelenting march against us.

And who are we to argue with a book that has been selling for over two thousand years?

It's not complicated. The unfortunate effects of global warming are mostly from energy; or more precisely, energy's use and how it travels from one place to another. And it goes like this. We burn fuels for their energy; they produce gaseous emissions, among which is carbon dioxide (CO_2); which in turn acts as an insulating blanket in our air. Radiative heat from the sun warms our planet, but less of it can escape back out due to this blanket, so we get warmer. And the more CO_2, the thicker the blanket, and the warmer we get.

Understanding global warming and its consequences is the first step in slowing it, reversing it, and finally defeating it.

CHAPTER CONTENTS

A Road Map to this Chapter

You know what a nice day for me is? It's when the next person who tells me "have a nice day" steps into an open manhole filled with alligators while cement is pouring into it. Now, there's the making of a really nice day. I meditate by clearing my mind of all thoughts except for that one image. Ah, nirvana.

It's the same with those socially challenged souls who tell me to "be safe." Yup, if they didn't tell me that, who knows what would happen. I might run headfirst into a brick wall or stick a wet finger into an electrical outlet. No telling what I might do without their timely instruction.

CO_2 is that kind of a person. Well-meaning and useful in small quantities but just so damned annoying when there's too much of it.

And it's the reason why temperatures are going up. Now understand, excess CO_2 doesn't increase our planet's temperature. Rather it increases its energy. So it is important we know the difference between the two.

That's the first thing covered: energy, how energy influences the heat content of our planet, and how that dictates its temperature.

For our home world, the energy balance, and therefore its resulting temperature, is due to three sources of heat radiation: the sun heating the earth, the earth reflecting some of that back out into space, and the earth emitting its own heat.

Think of cuddling on a sofa in front of a fireplace. Like the sun, the fireplace heats us up. And like the earth, we can feel the heat emanating from our partner. Put a screen between us and the fireplace and some of its heat will reflect away, cooling us down. Wrap ourselves in a blanket and we get hotter.

Which brings us back to CO_2. It is the blanket that keeps the heat from escaping. Add more blankets and we get hotter still. You'd think we would want to stop with piling on all these blankets, but maybe we are too distracted to notice. So more blankets, less heat escapes, and the hotter we get.

Where is this endless number of blankets coming from? Getting off the couch for a moment and back to reality, our CO_2 blankets come from burning fuels: gasoline, natural gas, oil, and the thickest blanket of all, coal.

There's more to it, but this is enough of a map to find our way, without stepping into any open manholes. So enjoy the rest of the chapter. Oh, and have a nice day.

Heat And Temperature—Similar But Different

Picture a tribe of cannibals preparing to cook an unfortunate clown who became separated from the circus. They put the clown into a big pot of water and light some wood placed beneath it. The energy from the burning wood goes into the water, which then increases its temperature. More wood is added, which adds more energy to the water, again increasing the amount of heat it has and again increasing its temperature further.

Cannibals aren't as backward as you might think. They know that to cook the clown, they need to raise his temperature, and to do that they need to raise the temperature of the water he is sitting in, and to do that they need to burn the wood to transfer its copious amounts of energy into the pot.

I suppose you can guess that the clown escaped unharmed. And you are probably going to think I will next write: "He escaped because cannibals never eat clowns. They taste funny."

Well, let me be clear on this point. I would never stoop that low. Besides, it's an old joke.

No, the clown escaped because cannibals can never agree on who gets the Funny Bone.

Anyway, clowns and cannibals aside, a more profitable way to look at this is from a water molecule's point of view. For that, we need to think small.

So do you suppose we can shrink down to a grain of sand the way Alice did when she drank the "Drink Me" potion. Not likely. But we can do a thought experiment as if we could.

Put a thermometer in that pot of boiling water. Next, pretend we can squeeze ourselves into the thermometer as if it were a submarine with a window looking into the steaming water. And let's further pretend our vision is so sharp and focused that we can see all the steam molecules individually moving, vibrating, and spinning.

The molecules moving here and there, hitting our window, hitting each other in an endless, random, violent dance. These molecules are constantly in motion. They never sleep, they never rest, and they are always moving. Some move faster than others, some are fat, some skinny, some are heavy, and some are light, but the one thing they all share is they are always moving. Imagine that—something that never stops.

When the molecules hit each other, one gains some energy and the other loses an equivalent amount of energy. Thus one speeds up and the other slows down. It's the same with shooting pool. When you hit the cue ball, you increase its speed. Then it collides with the eight ball, and the cue ball slows down and the eight ball speeds up.

Each molecule is moving on average at 750 miles per hour and 10 trillion will crash into our window each second, transferring some of their energy (and therefore their speed) to the glass wall, which in turn does the same to the mercury inside the thermometer. If the steam is heated, more molecules will strike the window harder, and this energy will transfer to the mercury in the thermometer. The mercury will then expand since its molecules speed up and therefore need more room to move. This change is read on the side of the thermometer as its temperature.

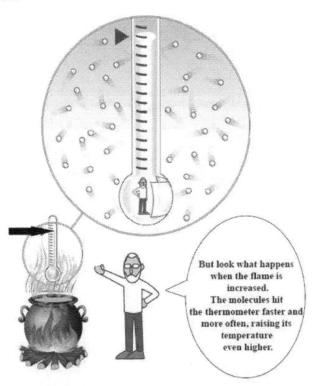

Adding more wood to the fire will increase the energy transfer to the water. The water gets hotter and the molecules move faster and bump into our window harder and more frequently. The faster and more frequently the molecules hit, the more they accelerate the molecules of the mercury in the thermometer, and the more the mercury will expand, indicating a higher temperature.

Some of the molecules move fast, some slow, some will hit the thermometer harder than others and therefore transfer a greater amount of energy. But since so many are hitting the window, the energy averages out, which brings me to the first key of this section:

Temperature is a measure of the average energy of the water.

Temperature is a measure of the average energy of the molecules (as determined by their speed[i]) in the water at some specific location, such as where the thermometer is placed in the above figures.

Place a finger in a cup water. You feel the water's temperature from the molecules colliding with your finger. The more molecules there are and the faster they go, the more your finger feels an increased force and therefore a higher temperature.

---

[i] More precisely, it is the molecules' kinetic energy which goes as their speed squared.

If you placed your finger in every possible location in the water and added the forces on your finger each time, that total would be the total heat content of the water.

Heat content is then the total amount of energy in the water, essentially the energy of every single molecule all added together.

In other words, heat content is the total energy the water contains, and temperature is its average energy at any point. Or to put it another way, temperature is the average speed of the molecules at one specific location. Heat content is the sum of the speeds of every molecule the water contains.

Temperature and Heat Are Different But Each Follows the Other

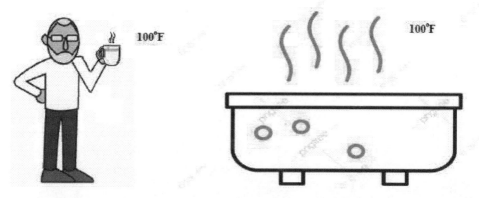

The hot water in my cup of tea and in the bathtub are at the same temperature so they have the same average energy. After all, that is what temperature is. But the bathtub contains much more heat energy than my cup of tea because it contains much more water which means it has many more moving molecules. And when you add up the energy of each of those bathtub water molecules, the sum is much greater than the energy of the fewer molecules in my cup of tea.

100°F   100°F

Heat content and temperature change together. If an object is heated (that is, more energy is transferred to it), the total energy and average energy of the molecules will go up, so both the heat content and temperature will increase.[ii] If the heat content of something increases, then its temperature must go up. Lie on the beach on a sunny day and you will get warm, but lie in the same place at night and you will get cold. The sunny day is increasing the energy you are absorbing so you get hot. At night you are radiating away your energy into space so you get cold.

Knowing heat content, energy transfer, and temperature is the first step in our campaign against global warming.[iii] And that's not something I would clown around about.

Just Like Diamonds, Energy Is Forever

Everything contains energy. There are no exceptions. Energy powers our factories, schools, homes, cars, even our bodies. A rock, inanimate, unmoving, unthinking, contains energy within the vibrations and movements of its molecules. For something we take for granted, energy is around us, within us—it *is* us. It is king.

And energy is forever and never goes away. It cannot be created or destroyed. It always is—almost godlike in its persistence. What the universe has provided is all there is. No more, no less.[iv] Energy can be transformed and moved about, and concentrated or diluted, but it never ceases to be.

This point is critical because it also means the energy of anything can always be accounted for and, more importantly, the temperature it produces. Looking at the flow of energy, whatever goes in one end must either come out the other end or increase the amount already present, which also increases its temperature. There are no other possibilities.

For instance, when you put a clown in a pot of boiling water, some of the heat from the wood escapes into the air, some of it heats the pot; and some of it heats the water raising its temperature; and some goes into the clown, also raising his temperature. Add up all these heat energies and they would be exactly the amount of heat energy the wood releases. Whatever the wood has to give must end up somewhere.

It's like a carwash. The number of cars going in must equal the number exiting plus any that remain inside. Cars don't enter and never emerge nor do cars magically materialize.

And what is true for a boiling pot of water is also true for our planet. If our planet is in harmony with all the energies going and coming from it, it will remain at a constant temperature. But if the harmony is broken, if the energy flows change, then the planet's temperature will go either up or down. It goes up if more energy is stored in it, and down if it loses some.

[ii] There are exceptions. For instance, when something melts or vaporizes, its temperature remains constant when more energy is added. Also, if two fluids at the same temperature are combined, the temperature remains unchanged but the total heat content will double. But on our trip through the global warming landscape we will not encounter these side roads, so we can dismiss them.

[iii] The words "heat" and "energy" are often used interchangeably. Heat is the thermal form of energy, whereas energy includes many other forms. In this book, the context will make it clear which I am referring to.

Energy and temperature are two of the keys that unlock our understanding of global warming.

---

iv Mass is a form of hidden energy, so in fact it is the combination of energy and mass that cannot be created or destroyed. This is clear from Einstein's famous $E = MC^2$, which states that energy (E) and mass (M) can be converted into each other. For instance, burning a cord of wood will necessarily sacrifice eight-billionths of an ounce of the wood. We can ignore it since the effect is so small.

"Here's an important Aside. While it is true energy always exists, there's a catch. Every time we use energy, its quality is diminished and becomes less useful the same way we all become less physically capable as we age. Concentrated energy does work for us, and after it completes its job, it becomes diluted, more spread out, and unable to do as much as it did the first time. Known as entropy, it describes an important limitation to the energy available to us."

"For instance, boil two gallons of water on a stove. Also roast a chicken in the oven. It will take the same energy to boil the water as is needed to roast the chicken. Now comes the catch. Drop the same uncooked chicken in the boiling water. Even though the water has the same amount of energy needed to roast the chicken (in fact, it could have much more, but it wouldn't matter), you can't use the boiling water to roast the chicken. The energy hasn't gone anywhere; it is still in the water, but it is more spread out. The water's temperature is only 212°F, but you need 350F to roast the chicken, so there is not enough drive in the water to get the meal done. The very best you could do is heat the chicken to a temperature approaching 212°F. You can boil as much water as you cared to, an entire ocean if you please, but you'd never be able to roast a single chicken with it. Even if you did it for an eternity.[v] Or even two eternities. It's not happening".

---

[v] Actually, you can do almost anything with an eternity in hand, so maybe you can roast that chicken if the water molecules arrange themselves in a highly improbable configuration in which the ones near the chicken randomly acquire an amount of energy much greater than the average. But let's not ruin a good story with facts. Especially improbable ones.

"The energy in the stove's flame that heated the oven to 350°F has a much higher quality than does the energy in the boiling water".

"Energy is useful. Entropy is fascinating".

Radiation And Its Various Colors Are An Important Key

In the previous examples, the flame needed molecules to heat the water. One molecule bumped into another and into the window of our imaginary submarine. But in the space between the sun and earth there are no molecules to speak of so there is nothing to carry its energy to us. Instead, the sun's energy finds a different way to move—it rides on electromagnetic waves, the way a surfer rides a wave toward the beach. Energy can travel through space without touching anything and without the need for anything material to carry it. This type of energy movement is called radiation.

While we experience radiation's energy every day (such as the sun heating us) but maybe without being aware, we also experience another important property, its wavelength. Visualize a wave, say the wave created when you pump the free end of a rope tied down at its other end. Its wavelength is the distance between two adjacent peaks. All waves behave like this.

But here's the thing—the shorter the wavelength, the more energy it contains. Pump the rope slowly and then quickly. Two things happen. First, the faster you pump, the more waves are generated and therefore the distance between two is smaller, meaning its wavelength is smaller. Second, you need more effort or energy to produce the faster waves, which means they have more energy than the slower waves. That is why gamma rays, the shortest, are deadly, since they pack such a big punch, whereas we are immersed in long wavelength radio waves all day without ever noticing.

After it rains, put the sun at your back and if you're lucky you will see a rainbow, which is the sun's light being separated into its component colors. This happens due to the water droplets suspended in the air. As the sun's light passes through the droplets, each of its colors, having a different wavelength, travels a different distance in the droplet, causing the colors to separate from each other leading to the beautiful colors we see.

It's clear these different colors of light behave much differently from each other. The red light moves a different distance in the water droplet than does the blue light, for instance, which must mean it differs from the blue light in some important aspect. That aspect is its wavelength and, consequently, its energy. And this makes a world of difference in not only how we see their colors but also how they act.

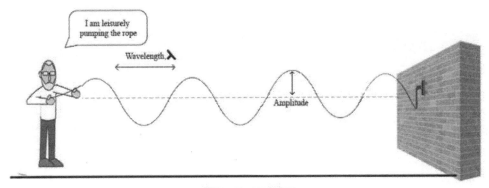

Transverse Wave

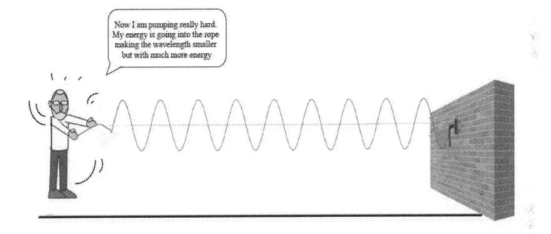

The following figure shows how this difference leads to significantly different properties of the different types of radiation. Shown is the entire radiation spectrum, including the visible (400 to 720 nanometers). In addition, there are microwaves to heat our food, radio waves to listen to our radio broadcasts, infrared to warm us, and so on, with the only distinction among them being their wavelength (and consequently their energy). Really, not much else but that is enough to make them behave radically differently from each other. Consider that we use x-rays, but not infrared, to see our bones, whereas infrared heats us up but x-rays won't.

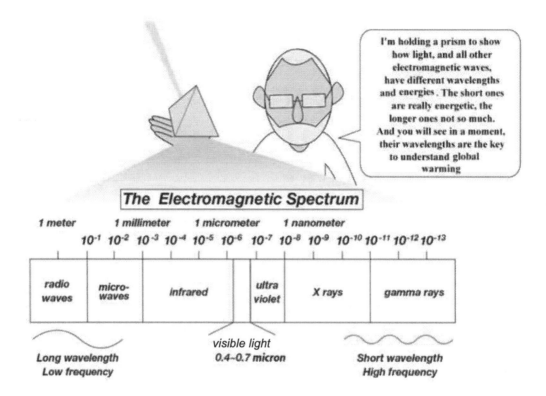

And this difference is crucially important for global warming. The following figure shows a comparison between the sun and the earth's emitted radiation. This is not the earth's reflected heat but rather the heat that it emits, the same way your snuggle buddy's body emits heat. We are already familiar with the sun's radiation as we see and feel it every day. The earth's radiation may be a new concept. Everything radiates to some degree, with the higher-temperature objects radiating much more than the lower ones.

Since the earth's temperature is thankfully much lower than the sun's temperature, it radiates much less energy and at much longer wavelengths. The sun radiates at a peak wavelength of about 0.5 microns whereas the earth's radiation is at a much longer wavelengths centered at about 15 microns.[vi]

The difference between the sun and earth's energy emissions is a key to understanding global warming. I'll leave you to ponder this for a bit before I return to it in the next section.

[vi] A micron is a millionth of a meter and is about the size of a single cell in your body. Now, that's really small.

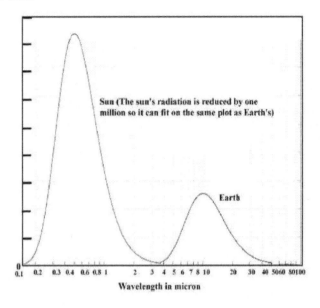

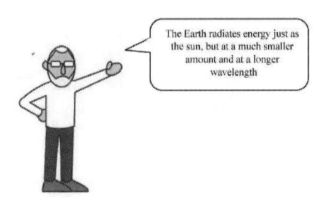

"*Here's an unsettling Aside. When a star explodes, called a supernova, it emits intense gamma radiation, which if close enough could destroy our ozone layer, creating havoc to all living things and maybe ending our short reign as the top dog in the food chain*".

"*So how close does a star have to be?" you might nervously ask. The kill zone is now thought to be fifty light-years away [1]. And while it might take as long as fifty years to reach our beautiful blue marble of a planet, we would never know about it since there is no way to detect the radiation until it hits us*".

"Don't do anything rash. The closest star that could go supernova is one hundred light-years away, well outside the kill zone, and it has a few million years before it pops".

"This brings me to gamma ray bursts, those pesky explosions are thought to be the most violent events in the entire universe and much stronger than wimpy supernovas. Their emitted radiation would also damage our ozone layer, but if close enough they could fry everything living on our planet. In a flash, all life on earth would cease faster than you could read this sentence".

"Again, don't go selling your belongings while shouting "Repent, the end is nigh." The likelihood is quite small as there aren't any potential gamma ray burst sources near us".

"Supernovas and gamma ray bursts illustrate the fascination of radiation. Where else can you think about global warming, instant planet annihilation, and cozy fireplaces, all from the comfort of your living room?"

"As a final Aside, radiation allows to see the past. Light moves at a blistering 186,000 miles per second. That's a lot of giddyup but it's not infinitely fast and when compared to the vastness of space, it can take some time to reach us. Take the closest star, our own sun. It is 93 million miles away so its light takes about eight minutes to reach us. Therefore we never see the sun as it really is, but what it was eight minutes ago. Now look at the stars at night, say the North Star, which is 323 light-years away. It takes 323 years for its light to reach us, so we are looking at something that happened over three centuries ago, right about the time the Mayflower landed at Plymouth. Peering deeper into space, we can see the farthest galaxies. When we look at their light through the James Webb orbiting telescope, we are watching the way it was 13.8 billion years ago, nearly at the beginning of the universe and time itself".

Add these to radiation's amazing feats.

How Earth Heats Up And Cools Down

Now back to the task at hand. Radiation heating and cooling determine the temperature of the earth; nothing else does. There are only three that matter: the sun's radiation heating the earth, the reflection of some of that heat back into space, and the earth radiating into space. See the following figure. Change any one of these three essential ingredients and the temperature will change. If the energy leaving the earth is reduced, the earth's energy content will increase and therefore its temperature will rise. No other outcome is possible.

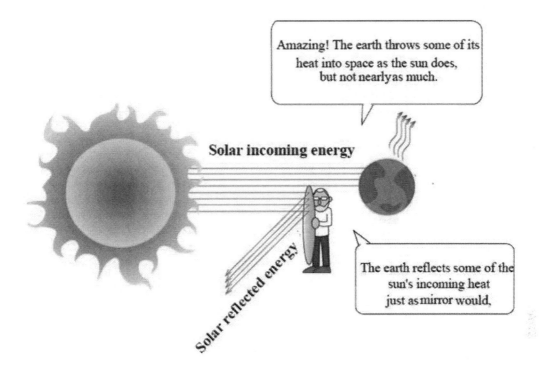

The sun's radiation has been either unchanging or going down for the past sixty years [2], so it has had no influence on the increase in our planet's energy content, which leaves the other two to consider.

And now for the coup de grace; in fact, the most crucial item: the radiation passing in either direction is influenced by the air and anything it contains, which acts as a brake on both. For instance, erupting volcanoes add ash to the air, which behaves like very tiny mirrors reflecting some of the sun's energy back into space. Haze can also play a role in earth's heat balance by blocking the sun's radiation reaching us. A study showed that haze contributed to a leveling off of the global warming heating affect from the 1950s to the 1970s before the heating resumed [3]. But greenhouse gases in the air will have the largest effect, overshadowing any haze or volcano produced particles.

Now we're getting somewhere. The next section is the final piece of the puzzle.

CO$_2$ Reduces Earth'S Cool Down, Leading To Rising Global Temperatures

Let's do another thought experiment. Imagine we have a one-way valve that allows heat to travel in one direction but hinders it from traveling in the opposite direction, just as a clogged drain slows down water from leaving a sink but doesn't change the water pouring from the faucet. If we were to put that valve on the radiation leaving and entering the earth, we could easily control its balance and therefore earth's temperature. For instance, if we orient the valve so that it reduces the energy leaving the earth but leaves the energy entering the earth unaffected, there would be an energy buildup on the earth, and its temperature would increase.

CO_2 molecule is such a one-way valve and has the regrettable property that it upsets the energy balance by inhibiting the energy loss that the earth radiates to space while leaving the heating energy from the sun unaffected.

It consists of an oxygen (O) atom in the middle of a line with a carbon (C) atom hanging on each end, forming a flexible molecule that can bend and stretch several ways when hit with radiative energy. The interesting thing is that the longer radiation wavelength in the infrared, which radiates outward from the earth, matches what is needed to vibrate the wings of the molecule, whereas the hotter and shorter wavelength radiation, coming from the sun, passes right through it without so much as a hello. CO_2 strongly absorbs radiation with a wavelength of 13 through 17 microns which, according to the above figure, is exactly the point that the earth radiates out and it is nowhere near the sun's radiation wavelength. That is to say, it blocks some of the earth's outgoing radiation but has no effect on the sun's incoming heat.

This then disrupts the earth's energy balance, increasing its heat content and therefore its temperature.

A familiar instance of this greenhouse effect is when your car is sitting in the summer sun with the windows rolled up. Like the atmosphere, the car's windows are mostly transparent to the sun's radiation, which warms the upholstery and the air inside the car. And just like the earth, the car seats in turn re-emit their heat as long wavelength radiation. But glass, like CO_2, blocks some of this outgoing radiation and thus traps the heat, and the car's interior gets hotter. That is why we should never leave children and pets inside a car alone.

A not so familiar instance is the runaway greenhouse effect on Venus. The Venusian atmosphere is mostly CO_2, much higher than the piddling amount we have here. And as a result, its surface is hot enough to melt lead.

Let's pretend you are an astronaut looking down at the earth. Further, let's pretend you have magic eyes and you can see, not just the visible part of the spectrum like we mortals but the entire spectrum. And finally, you can move back and forth through time. The following figure shows what you would see in the infrared part of the spectrum, 6 microns to 25 microns if you were looking down at the earth in the year 1760, right before the Industrial Revolution. The horizontal axis is the wavelength that separates the various spectrums, say red from green or say the visible from gamma rays. The vertical axis is the strength of the radiation, how much light you are seeing. You are looking at these beautiful colors, all with different hues and varying brightness. It's something no human has ever seen before.

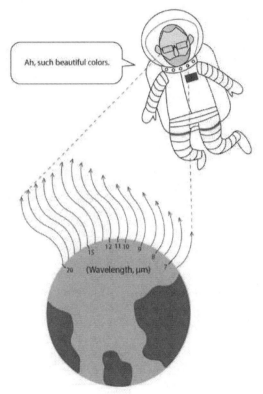

Fast-forward 260 years to today and the next figure is what you will see. These are measurements taken by the Nimbus 4 satellite while orbiting the earth 621 miles above Northern Africa in 1970 [4].

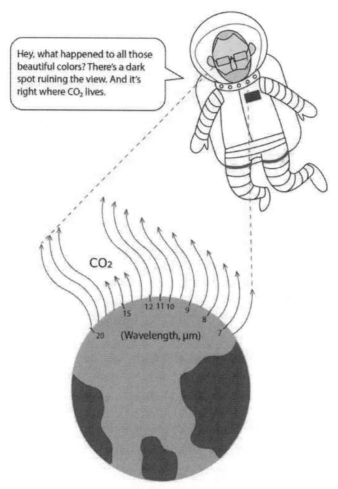

Something has changed. There is a dark spot in your vision, ruining the view, upending the beautiful, pristine colors from before. One is centered at 15 microns. No light, or at least much less light, is getting through. What's causing this dead spot? The 15-micron gap is exactly where CO_2 molecules absorb the outgoing radiation. This is a significant smoking gun if not the bullet itself. It is peer reviewed data collected by NASA that unambiguously shows CO_2 is blocking some of earth's radiation from escaping and acting as an insulator blanketing our world.

And the more CO_2 pumped into the air, the more of these one-way valves there are, and the greater the heating affect and the earth's temperature increase.

Not to beat a dead horse, but further proof is given in the next two figures which show that the amount of CO_2 in the air and the global average temperature move in lockstep. Remarkably, the first figure shows this connection going back

800,000 years and comes from ice core samples, which provide an accurate accounting of the atmosphere in the ancient past [5]. The figure shows more recent data from 1964 to 2008.[vii]

There is no question that increasing CO_2 will drive the earth to a higher temperature. It has already happened and it is continuing. But where does this CO_2 come from?

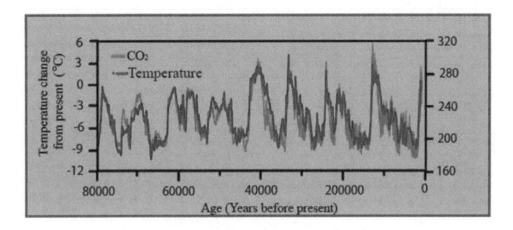

"Amazing! The temperature and CO_2 levels follow each other exactly for the past 80,000 years".

[vii] The vertical axis' ranges are designed to purposely superimpose the two curves for ease of viewing. For instance, CO_2 is shown going from 160 to 320 ppm rather than from 0 to 320.

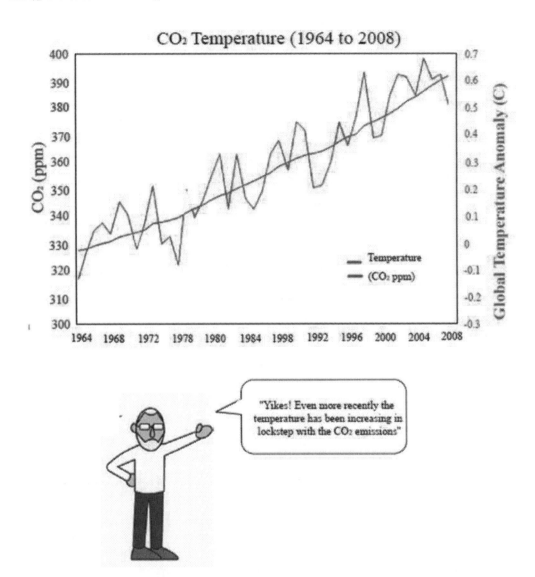

Fossil Fuels Provide Our Energy and, Unwittingly, CO₂ As Well

They were born millions of years ago in a place so extreme and dangerous you and I would not have lasted past lunch. Some of them weighed one hundred tons and others were the apex predators, the alpha males among giants. And now they are among us. Everywhere we go they are watching, waiting, infinitely patient, ready to act. They can hurl us across the sky and melt steel. There is no escaping them.

No, this is not the beginning of a cheesy sci-fi novel. It is the story of how our fuels were created and where CO_2 comes from. When trees, grass, plants, and animals, even dinosaurs, died and were buried, they were slowly converted into the fuels we use today; gasoline, coal, natural gas, and others, broadly called *fossil fuels*. Gravity crushed them under miles of rock, creating enormous pressures and temperatures over hundreds of millions of years, forcing their atoms and their bonds to rearrange into the energy powerhouses that we dig up and use today. Even the seemingly insignificant plankton, tiny organisms living in the sea, produced fossil fuels.

Despite the complexities of these myriad plants and creatures that make up our fuels, it is only two atoms that provide most of their energy: hydrogen (H) and carbon (C). The *Hindenburg* zeppelin disaster is a sad example of the energy locked up in the hydrogen atom. In 1937 the airship caught fire in Lakehurst, New Jersey, creating an immense fireball from the hydrogen used for buoyancy, killing most of the passengers and crew.

By themselves these two atoms are inert and will just sit there whiling away the time as if staring out the window in a nursing home. But introduce them to oxygen (O_2), of which there are bountiful amounts in the air, and they will throw down their walkers and light up like a roman candle. The C and H atoms chemically attach to the O_2 molecules, releasing their energy in the form of heat and visible flames. And this is what moves our cars and airplanes and heats our homes and food.

Like the cars in a carwash, the C atoms cannot disappear or morph into a different atom and because they combine with O_2, the resulting molecule (C atoms plus O_2 molecules) must be CO_2.[viii] Similarly, every four atoms of H combine with one molecule O_2 and form two molecules of water vapor (H_2O).

"As an Aside, everything burns, not just fuels. But the speed at which it occurs can be as slow as a crawling baby or as fast as a rocket. Leave a sliced apple on the counter and it will turn brown, evidence of oxygen combining with the organic atoms of the apple, slowly burning. I am burning right now as the energy I need to type this sentence is coming from oxygen combining with the food I ate. Fossil fuels burn the same way, just faster, producing a visible flame. And fastest of all is an explosion which is a flame moving at supersonic speeds".

Fuels produce CO_2 and some, like rabbits, are more prolific than others. The more C in a fuel, the more CO_2 will be generated when it is burned, and the more H in a fuel, the more water vapor there will be. Take natural gas, which many of us use to heat our homes. It has one C atom for every four H atoms and is written as CH_4. Coal, on the other hand, has more C atoms than H, so it produces more CO_2 than natural gas. For instance, anthracite coal has five C atoms for each four H atom.

But it gets worse. H produces more energy than does C, so for coal to produce the same amount of energy as natural gas, more of it needs to be used. Hence, in total, burning coal produces twice as much CO_2 as does natural gas.

The following table shows the CO_2 that each of the most widely used fuels produces for a given amount of energy (the second column) and how each fuel compares to natural gas (dividing the fuel's CO_2 by that of natural gas and shown in the third column).

A Comparison of the Emitted CO_2 of Some Fuels.

Fuel	CO_2 per Millions of Btu	CO_2 Compared to Natural Gas
Coal	229	2.0
Heating Oil	161	1.4
Diesel Fuel	161	1.4
Gasoline	157	1.3
Propane	139	1.2
Natural Gas	117	1
Hydrogen	0	0

[viii] A small amount of it (a few ppm) turns into CO, carbon monoxide, but mostly it turns into CO_2.

Hydrogen is an ideal fuel if you can get it, since it makes water vapor only when burned and creates no CO_2 by itself.

Coal is the worst of the lot, which is why it is thought of as the redheaded stepchild of fuels and why there is a push to replace it. Interestingly, coal happens to be on a steep decline in the U.S. but it has nothing to do with attempts to curb global warming. It is simply because natural gas is much cheaper to burn than coal which may be the first time in modern history. In the recent past, coal-fired power plants accounted for over 50 percent of our electric generation. In 2020, coal accounted for only 19 percent of our electric generation, dropping by more than half.

Coal's problems are obvious but we shouldn't forget that it has served us very well over the years. We built our economy on it. It has always been cheap, and although it needs to be phased out, it should be done with dignity and with thanks for a job well done.

CO_2 Is The Visiting Uncle Who Never Seems To Leave

Unfortunately, most of the CO_2 we emit will remain in the air from centuries to thousands of years since there is not enough growing vegetation, or other sinks to absorb it [6]. So for our lifespan and that of our descendants, we might as well accept that the CO_2 we create will be with us forever. And it accumulates, so as we burn more fuel, the amount in the air gets larger and larger with no end in sight. This is scary. It's an invading army whose numbers increase without bound.

Bad News—Earth's Reflectivity Is Going Down

As the sun's light passes through our atmosphere, the earth bounces some of it back into space, cooling the planet and acting like a global air conditioner. This is part of the earth's energy balance previously discussed. Ice and snow do a remarkable job of reflecting the sun's incoming energy. It's the same reason why we wear light-colored clothes in the summer as they reflect the sun's energy, whereas dark colored clothes soak up the rays.

But as our world heats up, the snow and ice melt, exposing more ground and oceans. For each patch of snow and ice we lose, we reduce this cooling effect by 90 percent. And of course, the more ice we lose, the higher our temperatures go and therefore the more ice we lose. A bit of a vicious cycle.

You can perform a simple experiment to confirm how much more energy ice reflects. Download a free light meter app onto your smartphone. I used

LightMeter, which is quite simple and easy to use. It measures the light in lumens, which is the amount of light it sees. On a sunny day take some ice and place it on the ground. Use your light meter to measure the light being reflected off of it. Next, take the same measurement on a piece of grass or ground nearby under the same sunny conditions. Be sure you use the same angle and distance for each one. I was maybe three inches from each and I put the ice completely in the view of the camera just off the vertical. Averaging five measurements, I got 157,648 lumens for the ice and 24,534 lumens for the grass. In this simple experiment, the ice reflected 84 percent more light than did the grass.

More Bad News—Melting Permafrost Will Release Alarming Amounts Of Greenhouse Gases

Permafrost is frozen ground located in the polar regions that sit over huge reservoirs of methane hydrates, which are greenhouse gases more potent than CO_2, The permafrost acts as a lid on this Pandora's box holding the hydrates in. As the permafrost thaws, the lid is removed and the methane becomes exposed and can escape into the air.

Permafrost covers 6.6 million square miles of the northern hemisphere and may be tens to hundreds of feet deep [7]. It is estimated that northern polar regions store twice the amount of carbon that is currently held in the atmosphere [8], about 1,700 billion tons. That's three times what we have emitted since the Industrial Revolution began [9].

As frozen soils thaw under warmer conditions, we are at serious risk of releasing some, and maybe quite a bit, of these greenhouse gases.

Just When You Thought It Couldn'T Get Any Worse

Still more bad news. Forests are burning in response to dry conditions from global warming which releases more CO_2 into the air, which is bad enough. But when the trees are gone, the remaining area may not regenerate, which removes an important source to soak up CO_2. Unfortunately, forests that are continually and intentionally cleared are gone forever. In 2019 there were over 140,000 fires in Brazil's rainforests, which will never regenerate [10]. The 2019 Australian wildfires emitted about 900 million tons of carbon into the air, nearly double what the entire continent does in a year [11]. Researchers attribute at least 30 percent of the Australian wildfires to global warming, and probably more [12].

But There Is Hope

At least one hopes there is hope, and I think there is. For every problem there lies an opportunity and global warming certainly falls into the problem category as much as anything our generation has faced. Part IV shows what we need to do, but please take heed. Time is running short.

Connecting The Dots

1. Energy cannot be created or destroyed (mostly). What goes in, minus what goes out, is what builds up. It's like overeating—here come the love handles.

2. As the heat (energy) of something goes up, its temperature must go up.

3. Earth's energy balance is determined by three thermal radiation sources: the sun's radiation heating the earth, the earth's radiation into space cooling the planet, and the earth reflecting some of the sun's radiation back into space, also cooling the planet.

4. The sun and the earth's radiation both pass through the air above our planet.

5. Burning fossil fuels create CO_2, a greenhouse gas that is deposited into our air.

6. CO_2 does not block the short wavelength radiation from the sun, but it impedes the long wavelength radiation leaving the earth.

7. This causes an imbalance in the earth's energy budget which results in an energy buildup over time as CO_2 increases.

8. This inexorably leads to a temperature increase.

9. Melting ice, thawing permafrost, and forest fires are making things much worse.

10. There is hope. We can beat global warming, but time is running out and we need to move quickly and decisively.

We can now answer the riddle posed at the chapter's beginning:

What is it that never gets smaller no matter how much is taken from it?

Energy. Energy is indestructible and no matter how many times it is used, it never diminishes. Yes, it becomes less useful and more spread out, but whatever was there to begin with remains.

Afterthought: "This Is The Way The World Ends, Not With A Bang But With A Whimper"—T. S. Eliot

Energy and mass combined are eternal. The universe started out with a certain amount which will never change for all time. The energy quality, however, is always degrading. It is like watching unfortunate immortals condemned to age but never die. Their numbers remain, but their health inexorably and continuously declines.

Which makes it easy to guess our universe's fate. The universe will continue to expand and cool, and the stars will use up their fuel until a point is reached when everything is at a uniform and low temperature. No useful energy will remain. No warming from the sun or any other stars will be possible. The energy will be dispersed throughout a much larger universe and it will be useless. And without useful energy, life anywhere will not be possible.

If you are one of the last survivors, you will be sitting on a barren cold rock of a planet and there will be nothing in the sky. No stars, no galaxies, no anything. They will have moved too far away to ever see again even if their stars hadn't already gone dark. In fact, at that point the universe will be expanding faster than light can travel, so even if something was still lit, its light would never reach the remaining people.

Some researchers believe there are other parallel universes adjacent to ours. The only way to continue living, it seems to me, is for those lonely survivors to either jump into one or at least draw energy from one of their stars.

Anyway it's not going to happen anytime soon, a few billion years at least. Stay tuned. I'll let you know if anything changes.

References

[1] Thompson A. How close would a supernova have to be to kill us all. Popular Mechanics 2017.

[2] Sun and Climate Moving in Opposite Directions. Available from: https://skepticalscience.com/solar-activity-sunspots-global-warming.htm/

[3] Voosen P. Ocean cycles sidelined in 20th century temperature record. Science 2019; 364(6443): 814.
[http://dx.doi.org/10.1126/science.364.6443.814] [PMID: 31147500]

[4] How Atmospheric Warming Works. ACS Climate Science Toolkit, ACS Chemistry for Life.

[5] "Temperature Change and Carbon Dioxide Change," NOAA, National Center for Environmental Information.

[6] Climate Change 2007, The Physical Science Basis. Intergovernmental Panel on Climate Change 2007; p. 25.

[7] Schuur T. The Permafrost Prediction. Sci Am 2016; 315(6): 56-61.
[http://dx.doi.org/10.1038/scientificamerican1216-56] [PMID: 28004690]

[8] "Digging Deeper: Permafrost and Climate Change," The Physical Environment.

[9] Hood M. Permafrost Collapse Is Speeding Climate Change: Study. France 2020.

[10] Ortiz E. Climate Change, Oxygen, and Biodiversity: Amazon Rainforest Fires Leave Plenty at Stake. NBC News 2019.

[11] Chow D. Australia Wildfires Unleash Millions of Tons of Carbon Dioxide. NBC News 2020.

[12] Fountain H. Climate Change Affected Australia's Wildfires, Scientists Confirm. New York Times 2020.

CHAPTER 5

Rising Temperatures are a Problem, but Not the Only Problem

"The world has moved on."

—*Roland Deschain, in The Dark Tower, by Stephen King*

CO_2 is insulating our world, leading to rising air temperatures. That much is certain. The IPCC has reviewed years of data and concluded that the world's average air temperature has already heated by 1.8°F (1°C) compared to preindustrial times. And they suggest we cease all greenhouse gas emissions by 2050 to keep the temperature rise to 2.7°F (1.5°C) or face serious life-changing consequences from the additional heat.

We might think that 2.7°F is lost in the noise, invisible in our daily weather that changes every day. We would be wrong. The amount of heat needed for such a rise is enormous,[i] and it represents a permanent increase over and above the weather fluctuations we see every day. It's like a rising tide that lifts the changing waves even higher than they would be otherwise. But it is also a tide that never retreats and keeps rising with no end in sight, covering the entire planet, not just a local beach. And it makes the extreme limits of the temperatures ever more extreme.

While bad enough, rising temperatures cascade into other problems making the outlook even more challenging. Because our world is intricately interconnected, one problem begets many others. An illness never affects one part of our body. A simple seasonal flu results in a fever, coughing, lethargy, weakness, and appetite loss. That's because, like our world, our body is also intricately interconnected. No one organ is isolated from the rest. Each depends on all the others to function. So it is with our planet.

Global warming has changed our world and it is never going back to the way it was. It has moved on. This is also certain. But by understanding these issues, we can arrest them and adapt. Understanding comes first. Then wisdom and then change. It is in our hands.

---

[i] It is enough energy to run the entire U.S. for the next six million years.

CHAPTER CONTENTS

Air and Ocean Temperatures are Increasing

The oceans have saved humanity more than you could ever know. They absorbed a monstrous right hook from global warming by capturing over 93 percent of its heat [1]. Think about that, what would have happened if more of that heat went into the air? The air temperature has already increased by 1.8°F,[ii] so how much further would it have gone if the oceans had been less generous? I'll leave that one locked up in my closet of anxieties.

There are good reasons for the oceans' largesse. First, oceans cover almost three-quarters of the earth's surface, so most of the sun's incoming radiation lands squarely on them. Second, the oceans act as a container for this flowing heat, as it can't escape out the bottom and therefore it remains captured. Third, the oceans' ability to hold heat is 1,000 times greater than the air because its mass is 250 times larger and also because water, pound for pound, can hold four times as much heat as can air. Finally, oceans can't radiate their heat back into space as well as land and ice, so they have an additional warming blanket covering them.

All in all, a lucky combination of ocean properties that have saved us. For now anyway.

But what the oceans take away they also give back. All of the heat they've been squirreling away is going to escape at some point and add to our already increased air temperature. Heat is not a homebody but loves to travel and will go downhill to something that is only slightly cooler than where it started from. So as the oceans absorb more heat, their temperature increases and at some point it will rise above the air's temperature and will gush out its heat with great enthusiasm.

It will not happen overnight because everything in life takes time. When it comes to moving heat around, the earth is downright lazy since it has considerable thermal inertia and it will be a while for its temperature to catch up to the heat imbalance. It's a delicate dance and it takes about forty years to finish, which is a blessing and a curse. A blessing because it delays the true damage we are taking and a curse because it masks the problem.

So, getting back to the IPCC's goal of limiting our temperature rise to 2.7°F; even if we achieve it, after forty years the oceans will add about another 1°F and we will be at a 3.8°F rise. No matter how you slice it, the earth's temperature will continue to increase at least for the next forty years and very likely will continue after that because we are not doing enough to eliminate greenhouse gases.

What will the new normal be like? Some local summer temperatures can be instructive as to what we will face. Summer temperatures will rise higher than the worldwide yearly average, so brace yourself. For instance, if we live in Milwaukee, our new summers will be the same as living in Florida, with a rise of 11°F. The following table shows a few other cities that, on average, will increase their summer temperatures by 10°F. Plus, peak temperatures will be much hotter than the average.

How Hot Will Your City Be if Current Emission Trends Continue?

Summers Now		Summers in 2100	
City	Average Temperature (°F)	City's New Location	Average Temperature (°F)
Bismarck, ND	81	Victoria, TX	93
Boston, MA	79	North Miami Beach, FL	89
Chicago, IL	82	Mesquite, TX	93
Dallas, TX	94	South of U.S. Border	104
Helena, MT	80	Riverside, CA	92
Memphis, TN	90	Laredo, TX	100
Milwaukee, WI	81	North Port, FL	92
New Orleans, LA	91	Pharr, TX	97
New York, NY	82	Lehigh Acres, FL	92
Omaha, NB	84	Harlingen, TX	95
Portland, ME	76	Newport News, VA	86
San Diego, CA	78	Lexington, KY	85
Seattle, WA	73	Placentia, CA	84
Washington, D.C.	87	Pharr, TX	97

Source: "1001 Blistering Future Summers"

By 2080 most U.S. cities will feel like they were relocated south by over five hundred miles [2].

Speed kills. So does temperature. When the temperature hits around 95°F to 105°F in humid weather or 115°F in dry weather, it can be deadly in just a few hours [3]. We aren't built for it. We are warm-blooded and we need to constantly cool ourselves. We do so by radiating out our heat and when that isn't enough, we perspire, which wets our skin and then evaporates, carrying away large amounts of our excess heat. Just think of the heat needed to boil water. That's the same amount our perspiration will carry away. An incredible machine. But it is limited and once you get past 95 to 115°F, our perspiration is not effective enough and we suffer from heat exhaustion leading to heat stroke, which is deadly if not quickly

---

[ii] This temperature rise, as defined by the IPCC, is the average air temperature at ground and sea level cross the globe as compared to the same average from 1850 to 1900.

attended to. The above table shows that of the fourteen cities listed, five will pass this threshold.

As temperatures rise, people's hearts are stressed more. Mortality in London increased by 3 percent for every 2°F increase when the temperature rose above 70°F [4]. And we may be heading to a global average temperature increase of more than 2.7°F and maybe as high as 5.4°F [5], which will make the above table even hotter.

What's worse, the oceans are getting tired of us. They have already absorbed 30 to 40 percent of the total CO_2 but that will gradually decrease as there is a limit to how much they can take in. There are only so many rooms at the inn. The air, on the other hand, is virtually unlimited to what it can take. The difference is that with air, adding a CO_2 molecule is like dropping a stone into the Grand Canyon. You can put as many in as you want. The gas molecules don't interact and they barely affect each other.

But the oceans, being a liquid, store CO_2 in a much different manner. They dissolve the CO_2 which causes it to chemically combine with the hydrogen atoms in water. But there is a limit to how many CO_2 molecules can be absorbed in this fashion. And we are now approaching that limit, so the oceans will start to absorb less CO_2 than in the past. And what they no longer absorb will, by default, go into the air. Woe unto us when that happens.

Looks like my closet of anxieties is biting back.

Storms are Getting Bigger and Meaner

Storms have always been destructive with some of them causing over $100 billion in damage [6]. But now we are unwisely feeding them energy, making them stronger.

Storms are powered by water and heat; water for their mass and heat for the energy that drives their speed and destructive forces. When storms form over the ocean, they draw in air containing large amounts of water vapor. As the storms push the air upward, the water vapor condenses, releasing large amounts of energy. It is the reverse of boiling water.

And global warming is supplying heat in copious amounts. The temperature and amount of water vapor are increasing as global warming temperatures shoot up, which feeds the beast. Remember, over 93 percent of the global warming heat is trapped in the oceans, which is not a good thing if you don't like unprecedented massive storms marching toward your town.

Plus warmer air can hold more water vapor. A 1.8°F increase, which we have already achieved and will soon pass, causes the air to hold 7 percent more moisture [7]. A 2.7°F increase adds over 10 percent moisture. More food for the beast.

Scientists predict that at least four more mega hurricanes per year will strike us by 2050 and these are not just your ordinary garden-variety blow your umbrella away type of storms. These are city busters, with winds between 130 to over 157 mph [8]. These are Hurricane Katrina type of storms, which killed more than eighteen hundred people and caused over $100 billion in damage [9]. Other researchers observed an increase of 25–30 percent in the number of category 4 and 5 hurricanes, the most destructive possible, for each 1.8°F increase in global average temperatures [10].

Haiyan, in 2013, was the strongest typhoon on record.[iii] What made it a superstar was unusually warm subsurface waters and rising sea levels which aided in its coastal assault [11]. Then in 2015 came Patricia which upended Haiyan's crown and became the strongest storm ever recorded [12].

Fantastic Beasts and where to find them. Coming soon to a town near you.

Don't Buy Beachfront Property

Florida Keys, New Orleans, Miami, lower Manhattan, Venice, London. Visit them while you can because they are going away. Not too far, but out of reach unless you have an air tank on your back. Before the end of the century, they will oftentimes flood because the sea is rising and its rise is accelerating. The following figure shows lower Manhattan with its predicted floodplain for 2100 on the right [13]. The light-colored areas show where flooding is likely. Lower Manhattan is only 7 to 13 feet above the sea level so even a slight sea level rise will have this effect.

Global warming has two ways to raise sea levels. The first, as water heats up it expands like any other material. You use this fact when you run hot water over a stubborn metal lid. Metal expands more than the glass jar it is attached to, so it becomes loose enough to remove. Similarly, as the sea expands, it becomes more voluminous and its level must go up. Why does water expand when heated? As the water molecules' temperature increases, they travel faster and vibrate more, pushing adjacent water molecules farther away. This leads to an increase in volume because the molecules are taking up more space.

[iii] Hurricanes, typhoons, and cyclones are different names for storms that are used in different parts of the world. In the Atlantic and Northeast Pacific, storms are called hurricanes; in the North West Pacific, storms are called typhoons; and in the South Pacific and Indian Ocean, they are called cyclones.

**Global Warming Will Cause Extensive Flooding in Manhattan
Before (Left) and After (Right)**

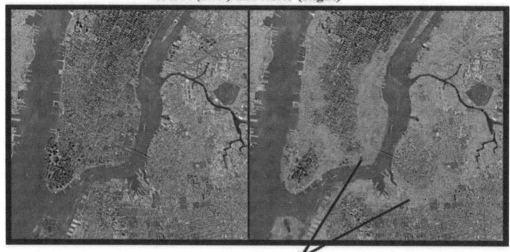

Land Loss due to Global Warming Flooding

Global warming's second effect is to melt glaciers and ice sheets causing liquid water to run into the oceans raising their level no different than topping off your drink. Plus this melting may also unblock solid ice which will then slide into the sea, once again raising its level.

Greenland, the Arctic, and the Antarctic all contain large amounts of ice and their temperatures are rising five times faster than the global average [14]. Greenland lost 662 billion tons of ice in 2019. If all of Greenland's ice melted entirely, it would raise the world's sea level by 23 feet [15]. If the Antarctic lost all of its ice, it would permanently swamp every coastal city. Arctic sea ice is disappearing at such a rapid rate that in the next thirty years it may be ice-free all summer [16]. No one, either living or dead, has ever seen anything like this.

Sea levels' rise is accelerating. The following figure shows actual sea level increases from 1900 to after 2020 [17]. The World Meteorological Organization reports that from 1993 it was rising at 1.3 inches per year but from May 2014 to 2019 it increased to 2.0 inches per year [18].

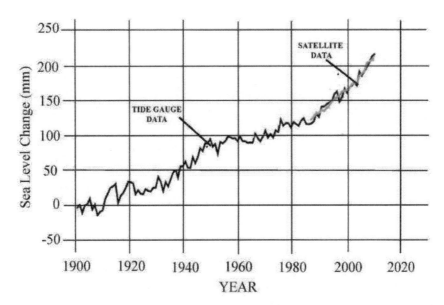

And by 2100, seas will rise from one foot to almost four feet [19], whereas the National Oceanic and Atmospheric Administration says seas may rise by as much as 8.2 feet. That would be a 464-foot march inland and maybe much more, with storm surges taking it even farther. And these are considered conservative so it could be quite higher.

The East and Gulf coasts are seeing frequent flooding and in the next thirty years, three hundred thousand coastal homes worth $117.5 billion will see chronic flooding.

But the real problem with rising sea levels is not the sea levels themselves but how they empower the already destructive forces of global warming–enhanced storms. When coastal oceans are higher, storms can push the punishing water surges farther inland and will also make them more intense.

I can see an objection rising in your mind about melting ice. *"Ice cubes that melt in my drink never cause the drink to overflow so how can melting ice raise sea levels?"* Good point and ice that floats in the sea don't raise sea levels when it melts, at least not by much. But that ice is not the problem. It's the ice sitting on land that, when it melts, flows into the ocean raising its level. Further, ice that was formally blocked by the ice that just melted may now slip into the ocean, further raising sea levels.

"Which brings me to an interesting Aside having to do with this very odd property of water and ice without which you would not be reading this (and I would never

have written it). In contrast to most other materials, ice floats because it is less dense than liquid water. Most solid materials are heavier than their liquids. Scrap steel, for instance, will sink to the bottom of a ladle when thrown on top of molten steel".

"But consider if ice and water behaved as most other materials and ice sank in the water. Sure, cold drinks everywhere would be overflowing, but that would be the least of our problems. Anytime some piece of ocean froze it would sink to the bottom and would then be under enormous pressure from the seawater sitting on top of it. Under that pressure and being so far away from sunlight, it would never remelt. So every time a slice of water froze it would sink to the bottom of the ocean and could never unfreeze. What would happen? A piece here would freeze than another there, and eventually, most of the oceans would be frozen solid and nothing could ever reverse it".

"This is not a good thing if you were trying to kick-start life millions of years ago. The first animals started in the oceans and crawled onto land, but they would never have existed and they would never have done their now famous crawl if ice behaved as most other materials. Where would that have left us? Nowhere. Without them, we would never have been".

"So in a backhanded sort of way, the fact that I am writing this is proof ice floats".

"Now to conclude my somewhat wordy Aside, I tip my hat to the ice/water inventor, wherever He or She is".

Diseases will Get Worse

Many diseases are caused by parasites and viruses hitchhiking inside mosquitoes which are then surgically injected into our blood. Like heat-seeking missiles, mosquitoes use our body heat, sweat, and exhaled CO_2 to zero in on us. These little death machines kill 2 million people globally each year. Over the past two hundred thousand years, they have killed 52 billion people out of 108 billion people who have ever existed [20]. That's almost half of humanity. Nothing else has ever come close.

There are 110 trillion mosquitoes in the world which makes me wonder how one goes about counting them, especially since they live for only seven days. Quickly, I guess. But assuming it is reasonably accurate, then they outnumber us by fourteen thousand to one.

And, yes, it gets worse. Mosquitoes thrive in warm, moist weather and flooding, all of which are brought to us by global warming. The warmer it gets and the longer it stays warm, the more mosquitoes will be born and the longer they will survive, giving them more opportunities to infect us with whatever disease of the day they are carrying.

Put yourself in their shoes. They need to cram a lifetime into seven days. No sleeping in for them. So they will jump at any opening to extend their life and put more of their kin into play.

Diseases from mosquitoes, ticks, and fleas have tripled between 2014 and 2016, with over 640,000 illness reported in the U.S [21]. Maine alone had a twenty-fold increase in Lyme disease illnesses. And now Lyme disease has been recorded in all fifty states.

Thawing permafrost, that bugaboo from chapter 4, has another trick to use against us. In Siberia, thawed permafrost released anthrax into nearby water and soil and into the town's food [22]. It killed one boy, sickened about 120 others, and killed 2,300 reindeer.

What other tiny monsters are waiting to be released from the thawing permafrost? Maybe none, as one researcher claims that anthrax periodically infects people and the permafrost melting may have only been a side consequence. Also, pathogens that have been dormant in the permafrost are likely adapted to cold deep soil and may not thrive in warm human bodies [23]. Still, I'd rather keep an open mind on this one, since there is too much we don't know about it.

Global warming will throw other health problems our way. A paper in Lancet, a prestigious medical journal states, "Climate change is the biggest global health threat in the 21st century . . ., and it threatens to undermine the last half century of gains in development and global health [24]".

The trend is clear. Global warming is causing diseases and many other health problems which gives us another reason, not that we needed one, to stop greenhouse gas emissions.

Food will be Harder to Find

People, ever a problem for all that ails us, are at it again. In 1990 we had 1.6 billion people, a number our planet could comfortably handle. In 2020, that went to 7.8 billion, tough but still holding it together. Now those 7.8 billion people want to raise the stakes and push our population to 9 billion by 2050, a 15 percent increase. More food will be needed and not just another 15 percent. That would be too easy because people don't just want food, they want high-calorie, tasty food that comes with an improved standard of living. So we will need to increase food production by 50 percent in order to feed these new 1.2 billion people [25].

That's tough enough but global warming is going to make it even harder. Rising temperatures will lead to a 2 percent reduction in crops each decade for the rest of this century [26]. Soybean and corn crops drastically decline when the temperature climbs to around 84°F. According to Wolfram Schlenker, plants may be hardwired to shut down at temperatures above this [27].

Insects, our nemesis from the previous section, aren't content with just infecting us with diseases. They also take a swipe at our food. They already consume 5–20 percent of major grain crops, which will increase another 10–25 percent per 2°F of warming for wheat, rice, and corn [28]. At 3.6°F, wheat will lose an additional 65 million tons of wheat per year, enough to feed 65 million people.

Warmer weather provides an encouraging habitat for pests, increasing their numbers but warmer temperatures also increase their metabolism making them hungrier. And it is worse in temperate regions which include most of the U.S. and, most importantly, our Midwest breadbasket where most of our wheat and corn are grown.

Here's another issue. Bite down on a cut lemon and suck in its sour juices. I can picture your grimacing face. Now you know how the oceans feel. The oceans' acidity increased 26 percent over the preindustrial era because they are absorbing about 22 million tons of CO_2 every day [29]. The oceans are acidifying one hundred times faster than at any time in the last two hundred thousand years and maybe in all of Earth's history. The CO_2 reacts with water to form carbonic acid which eats away at oyster and clam shells, leading to their demise [30].

Further, warming oceans mean they can hold less oxygen which is critical for fish. Between 1960 and 2010 the amount of oxygen dissolved in the oceans declined by 2 percent on average and as much as 40 percent in some tropical locations [31]. This is good for jellyfish but not good for tuna and other fish we like to eat.

Nothing escapes global warming.

Plants and Animals are Disappearing in Droves

What would it be like if you woke up in a strange house with people you didn't know speaking a language you didn't understand? With different laws, an incomprehensible culture, two moons, where up was down, and anchovies tasted sweet? Could you prosper? Could you even survive?

That is what animals have awoken to. To the bewildering changes global warming has made to their land, their food, their enemies, their lives. Their world, always dangerous, was at least understandable and stable. And when changes occurred, they had time to adapt and could be sure their children would survive after they were gone. Overnight, global warming washed that away and now they can only try to stay ahead of the heat, the drought, the changed landscape, the new predators, the diminished range, the scarce food. They can't console one another or cry or complain or understand or blame. That's not their way. They can only hope they can do enough, maybe just enough, for their children to have a shot at producing the next generation. Many will not.

There is unfortunate fossil evidence to support these fears. In the Great Dying, 252 million years ago, 70 percent of life on land and 95 percent in the oceans went extinct and the smoking gun was large amounts of CO_2 emitted from Siberian volcanoes [32]. No, this wasn't from SUVs, but the cause, CO_2, was the same and we should take heed of its warning to us.[iv]

And now we are in the middle of the sixth mass extinction in our planet's history [33]. Only five times before have so many species and so much biodiversity been lost so quickly.

In a detective story worthy of Hercule Poirot, scientists have determined how many species we have killed. From the fossil records they can count the number of extinct species over millions of years which tells them what the background extinction rate should be if we were minding our own business. Look around you. Does it look like we have been minding our own business? No, and now current extinction rates are a thousand times greater than they should be, and up to 1 million plants and animals will go extinct that never should have [34]. It's not just global warming causing the havoc; habitat destruction, over fishing, and an ever-increasing human population are adding their share of mayhem. But global warming plays a major role.

In ten years we will lose one-quarter of all the insects, in fifty years half will be gone, and in one hundred years all will be gone. Not good. No doubt some insects are pests but many do our bidding. They pollinate fruits, vegetables, and nuts. They also sit on a lower rung of the food chain and are eaten by fish, birds, and mammals that we then eat, so their demise percolates up the food chain all the way to us. A reduction in this many species could cause an entire community of animals and plants that rely on them to collapse.

Bumblebees' decline is an ominous sign especially since they are the top agricultural pollinators and are crucial for crops such as tomatoes, berries, melons, apples, broccoli, squash, and almonds [35]. Habitat destruction and pesticide

---

[iv] As an Aside, this apocalypse cleared the way for dinosaurs' record breaking 167 million years of domination and perhaps, as a statistical anomaly, led to us when an asteroid ended their long reign.

poisoning are killing them. But global warming has its hand in it as well. Bees prefer cool, wet conditions. Global warming is drying out their territory and increasing its temperature, so bees could be extinct in only a few decades [36]. Reduction in bees and other pollinators will destroy $577 billion of crops.

Extinct species are just the tip of the iceberg. There are other not-yet-dead species waiting their turn, heading inexorably into oblivion. These are the species in which whole groups of their populations have disappeared or their range has been so greatly diminished they cannot support their numbers and are fast heading to extinction [37]. They may not be there yet, but their destiny is sealed. Nothing can reverse it. And population decimations are many times larger than the species counted as extinct, which means there are considerably many more species in the extinction pipeline. Gerardo Ceballos stated there are 515 species he knows of with less than one thousand individuals [38], which implies there is no way to save them.

We are the top dog in the food chain. Everything else below serves us from tuna to cows to chickens to bumblebees. This is nature's hierarchical rule of power. Those at the top use those below. But this is not always a good thing, because any disruption anywhere in this pyramid will always affect us. Reduce biodiversity and we can only be worse off. No scenario that anyone has thought of has us benefiting. That's the disadvantage of being on top. Everything serves us, so any loss, no matter how small, is our loss.

Three-quarters of all animals may be gone and are never coming back. And if three-quarters of them can go, then why not all of them, and if all of them, then why not us? Another study found that 40 percent of all primates are at risk of extinction. Primates are monkeys, gorillas, apes, oh and yes, people too. No one ever said we are exempt.

Changes are Occurring Quicker than any Time in Earth's History

Global warming is fundamentally a problem of change and speed. How fast the changes occur tells us how much time we have to adapt. Slow change, long times are good. Fast change, short times are bad. It is one thing to fix a slow drip but quite another when the pipe bursts.

The following figure shows how CO_2 has changed over the past 800,000 years [39]. Note the spike right after the Industrial Revolution. According to one study, global warming is occurring at a pace hundreds to thousands times faster than at any time in the last 65 million years [40]. The speed at which it has changed is unprecedented in the earth's long geological history, maybe 4.5 billion years' worth. Global warming is speeding along like a Bugatti race car.

Carbon Dioxide Levels have Skyrocketed in Modern Times

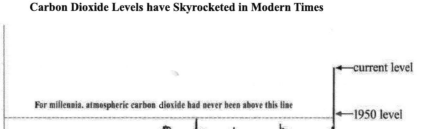

Some supporting facts:

• The Paleocene-Eocene thermal maximum increased by 9°F in about 10,000 years due to volcanic CO_2 release [41]. That's 1.8°F every 2,000 years. No contest. We are doing the same in decades.

• After the last ice age, it took 12,500 years for the global temperature to rise by 23°F, which is almost 2°F every 1,000 years. Today the temperature has already increased by that amount in only 50 years which is twenty times faster.

• It took 50,000 to 2.76 million years for each of the previous five mass extinctions to occur [42]. We are now only 100 years into global warming and are well on our way to handily beating those times.

These are races we don't want to win.

The pipe is bursting and for animals, trees, plants, insects, and flowers, it is not going well. Before the Industrial Revolution, they readily adapted to environmental changes. Animals can move their foraging range farther north if temperatures increase. Even trees and plants can move. No, not like the Ents in *Lord of the Rings*. Trees cannot uproot and leisurely walk to a better neighborhood. Instead, they disperse their seeds and some will fall or be carried in a more favorable direction and these saplings will have a better chance of surviving. They do the same when it's their turn, which then moves the tree line in an overall better direction. In this fashion, trees can migrate about 0.1 miles per year on average [43].

But for plants and trees to survive they need to move 62 to 93 miles for every 1.8°F rise. The math doesn't work for them. We have pushed the 1.8°F increase over about fifty years so that means trees need to migrate 1 to 2 miles per year, well short of the 0.1 miles per year they are capable of.

Increased Violence is Inevitable

Conflict over scarce resources may be the second oldest profession and can be traced at least to the beginning of recorded history. In 2500 BC a water dispute arose in Lagash-Umma, which is in modern-day Iraq. More recently, the Japanese attacked Pearl Harbor because President Roosevelt cut off their oil and for the same reason and in the same war, Germany invaded Russia. The U.S. was embroiled in the Gulf War to protect its oil interests. Cattle ranchers have clashed with farmers for control over the land they both needed. In 2020, poachers killed National Park rangers in Virunga National Park in the Democratic Republic of Congo. The 1994 Rwanda genocide was due in part to a population increase in sub-Saharan Africa that put excessive pressure on land and freshwater use [44].

And now we have the perfect storm. A rapidly increasing population never before seen on planet earth; alarming loss of habitat; cheap weapons aplenty; and global warming driving food scarcity, water shortages, flooding, and disease. And let's throw in nationalistic jingo for good measure. What could go wrong?

A lot, as the IPCC discusses in their 2014 "Impacts, Adaptation, and Vulnerability" report [45]. Global warming can bring about violent conflict by amplifying existing issues. As global warming stress increases, conflict is more likely when populations are insecure due to migration, income reductions, forced migrations, and adverse cultural changes. Global warming may not directly contribute to violence, but it will certainly hasten triggers already present. It sure isn't going to help any.

Further, David Rotman shows a 2°C (3.6°F) rise increases the probability of violence by 7 percent per year [46].

One study showed that higher temperatures generally predict more civil wars and predicts a 54 percent increase in wars in Africa by 2030 [47]. Civil wars doubled in the tropics during high temperatures and in sub-Saharan Africa conflicts increased by 30 percent for every 1.8°F increase [48].

Nils Gleditsch is more optimistic and first cites the reduction in battlefield deaths since the end of World War II [49] which went from three hundred thousand in the decade immediately after and then down to forty-four thousand in the first decade of this century. He next reviews the IPCC documents on global warming and violence and concludes there is little evidence connecting the two.

His welcomed conclusion is that global warming will not increase violence, or perhaps a better interpretation of his work should be—it has not *yet* increased violence. His assessment may be correct in so far as violence and global warming have gone about their business to date. Where I am concerned, and what he could not address, is what happens in the future when global warming becomes more extreme. We don't yet have our backs against a wall, so there is no need for our ancient phobias to kick in. So far. But if global temperatures are not held in check at 2.7°F, and if we pass tipping points of one kind or another, all bets are off.

Violence will also affect us individually. By 2050 warmer weather could boost interpersonal violence by 16 percent and group conflicts in some regions by 50 percent [50]. People become more violent as it gets hot, and urban homicides increase. Also, higher temperatures may affect a person's ability to reason. In New York City, for example, a 5.5°F increase over average temperature led to a 4 percent increase in personal violence and a 14 percent increase in group violence [51].

But Living has Always been Hazardous to our Health

All these problems are serious and threaten our well-being. There is no way to sugarcoat it nor should we. But when have we ever lived in a safe, stable world without threats? The Pax Romana—the two hundred years of world peace during the height of the Roman Empire? Maybe. But that was two thousand years ago, too distant for me to recall.

The mere fact that we are alive means threats will always be present so we hardly live in unusual times. Remember, our grandparents and their grandparents lived through the Spanish flu, in which 50 million people died; the horrors of two world wars, in which 95 million people died and whole cities and countries destroyed; and the Great Depression, in which the stock market lost 90 percent of its value and eleven thousand banks failed. More recently we have been through 9/11 and the COVID-19 pandemic. We've been there before, we are there now, and we will be there again.

Yet still we stand. And we have not just survived, we have thrived. That is why we are at the top of the food chain and why we will remain so. It's not the problems we face but how we face them. So let's get about it shall we. Time is running short.

Connecting the Dots

1. Temperatures are increasing. By a lot. But greenhouse gases don't just stop at increasing our temperatures. We should be so lucky.

2. The oceans have been absorbing most of the excess heat due to global warming, which is a good thing. For the moment anyway. But their temperature has been going up and in about forty years they will start sending some of its heat our way, further heating the air. That is why, no matter what we do, we are committed to an additional temperature rise of at least 1°F.

3. Oceans, ever our friend, have also been absorbing much of the CO_2 we are admitting. But they are reaching their limit, and when they can no longer hold any more, all of it will go into the air, making a bad situation worse.

4. Global warming is heating the oceans to unprecedented levels, which is exactly what storms love. Storms are powered by ocean water and the hotter the water, the more massive and destructive are the storms. And they are heading our way.

5. The oceans are rising. As they heat up they expand which makes them rise. Plus, as glaciers melt they runoff into the oceans making them rise even more. Not worried enough yet? Here's one more. The rising oceans exacerbate the effect of storms, pushing flooding deeper into our cities with more destructive winds.

6. Mosquitoes thrive in hot humid weather and they adore flooding, of which plenty are coming. And they love sticking their proboscises into human flesh and injecting viruses and other nasty disease-causing parasites. Their numbers are increasing and will continue to do so unless we put the brakes on global warming.

7. Rising temperatures will also reduce crops. Plants slow down and are hardwired to shut down completely if it gets too hot for them. Hot? Remember hot? That's what happens with global warming.

8. We are now in the midst of a mass extinction seen only five times throughout our planet's history. Many animals and plants are disappearing and they are never coming back. That affects us dearly. We rely on animals and plants for food of course, but the intricate web of interconnected animals and plants, which we evolved to thrive in, is being disrupted. No good will come of it for us.

9. Global warming change is occurring faster than plants and animals can adjust to. Humans too. It is happening at a pace hundreds to thousands times faster than at any other time in Earth's history. Global warming, difficult by itself, has now entered into the realm of nearly impossible.

10. Rising temperatures decrease our food, water, habitable areas, and more. Nations have fought over less. Increased violence is inevitable.

11. However, we are not helpless. And we have faced threats throughout our existence. Global warming is just one more. But we need to act. And act fast. There is a way. But let's not dawdle and make global warming our final challenge.

Afterthought: can Global Warming cause Cold Weather?

This seems like a preposterous contradiction, and the deniers are not going to like it, but yes it can. Blame the jet stream, a ribbon of air circling the northern hemisphere at high altitude and flowing easterly at 250 miles per hour. In one sense, you can think of it as a fence separating the northern cold weather from the southern hot weather. But because the Arctic is heating up faster than the mid-latitudes, it produces undulating wiggles in the jet stream and when one of the wiggles moves south, the cold arctic air follows it resulting in unusually local cold weather. This wiggle anomaly was responsible for record snowfalls in New York in 2015.

I like this one because it is true and it tweaks the deniers' noses, which if Pinocchio is any guide, there is a lot of nose to tweak. I can probably do the tweaking sitting across town from them.

One might say these are cold facts freezing out the deniers' position and snowing on their parade while giving them a frosty temperament since their ability to absorb correct ideas is glacially slow. One might, but I never would.

References

[1]   Upton J. Climate Central, "Rapidly Warming Oceans Set to Release Heat into the Atmosphere,". Sci Am 2014; 30.

[2]   Chodosh S. Here's How Global Warming Will Change Your Town's Weather by 2080 2019. Available from: https://www.popsci.com/2080-american-towns-climate/

[3]   aMatthews T. Global Warming Now Pushing Heat into Territory Humans Cannot Tolerate. Yahoo News 2020.bWallace D. The Uninhabitable Earth. New York Times 2017.cMunroe R. How Hot Is Too Hot?. New York Times 2020.

[4]   Dalton C. When Temperatures Rise, So Do Health Problems. NPR 2019.

[5]   2019 Concludes a Decade of Exceptional Global Heat and High-Impact Weather. World Metrological Organization 2019.

[6]   Berardelli J. How Climate Change Is Making Hurricanes More Dangerous. Yale Climate Connections 2019.

[7]   O'Mally I. Climate Change Has Altered Where Tropical Cyclones Occur, Study Says. Weather Network 2020.

[8]   Hertsgaard M. Hot, Living through the Next Fifty Years on Earth. Boston: Houghton Mifflin Harcourt 2011.

[9]   Zimmermann K. Hurricane Katrina, Facts, Damage, & Aftermath. Live Science 2015.

[10]   Berardelli J. How Climate Change Is Making Hurricanes More Dangerous. Yale Climate Connections 2019.

[11]   Clues to Supertyphoon's Ferocity Found in the Western Pacific. Science 2013; 342: 29.

[12]   Botelho G. Mexico Hunkers Down to Patricia, 'The Most Dangerous Storm in History. CNN 2015.

[13]   NYC Flood Hazard Mapper.

[14]   Qiu J. Winds of Change. Science 2012; 338(6109): 879-81.
        [http://dx.doi.org/10.1126/science.338.6109.879] [PMID: 23161970]

[15]   Abnett K. Greenland Ice Sheets Shrink by Record Amount—Climate Study. Reuters 2020.

[16]   Rice D. Arctic Will See Ice-Free Summers by 2050 as Globe Warms, Study Says. USA Today 2020.

[17]   Global Climate Change. NASA Available from: https://climate.nasa.gov/vital-signs/sea-level/

[18]   McGrath M. Climate Change: Impacts 'Accelerating' as Leaders Gather for UN Talks 2019.

[19]   Moore R. IPCC Report: Sea Level Rise Is a Present and Future Danger. BBC 2019.

[20]   Dodwell D. Climate Change Isn't Bad for Everyone—Disease-Carrying Mosquitoes May Be Heading for a Heyday. Outside In 2019.

[21]   Galea S. Feeling Sick? You May Have a Case of Climate Change. 2018.

[22]   Goudarzi S. What Lies Beneath. Sci Am 2016; 315(5): 11-2.
        [http://dx.doi.org/10.1038/scientificamerican1116-11] [PMID: 27918518]

[23]   Doucleff M. Are There Zombie Viruses—Like the 1918 Flu—Thawing in the Permafrost?. NPR 2020.

[24]   Parshley L. Catching Fever. Sci Am 2018; 318(5): 58-65.
        [http://dx.doi.org/10.1038/scientificamerican0518-58] [PMID: 29672493]

[25]   Mbow CC, *et al.* "2019: Food Security," in Climate Change and Land: An IPCC Special Report on Climate Change, Desertification, Land Degradation, Sustainable Land Management, Food Security, and Greenhouse Gas Fluxes in Terrestrial Ecosystems, Intergovernmental Panel on Climate Change.

[26]   "Climate Change Seen Posing Risk to Food Supplies," New York Times, 3 November 2013.

[27]   Cited in Rotman, D., "Climate Change Will Make It Increasingly Difficult to Feed the World," MIT Technology Review, vol. 117, no 1.

[28]   Deutsch CA, Tewksbury JJ, Tigchelaar M, *et al.* Increase in crop losses to insect pests in a warming climate. Science 2018; 361(6405): 916-9.
[http://dx.doi.org/10.1126/science.aat3466] [PMID: 30166490]

[29]   2019 Concludes a Decade of Exceptional Global Heat and High-Impact Weather. World Metrological Organization 2019.

[30]   NCEL Fact Sheet: Ocean Acidification," National Caucus of Environmental Legislators.

[31]   McGrath M. Climate Change: Oceans Running Out of Oxygen as Temperatures Rise. BBC 2019.

[32]   Radford T. "Ocean Acidification Triggered Mass Extinctions 252 Million Years Ago," RTCC, 18 April 2015; Lucas, J., "Oxygen Depletion Smothered Marine Life in Earth's Largest Mass Extinction,". Sci Am 2018.

[33]   Vital Variation. Science 2013; 339: 11.

[34]   Woodward A. 18 Signs We're in the Middle of a 6th Mass Extinction. Business Insider 2019.

[35]   Hernandez D. The Earth's Sixth Mass Extinction Is Accelerating. Popular Mechanics 2020.

[36]   Slater G. Scientists Warn Bumblebees Might Be Going Extinct Due to Climate Crisis. Sci Am 2020.

[37]   Ceballos G, Ehrlich PR, Dirzo R. Biological annihilation via the ongoing sixth mass extinction signaled by vertebrate population losses and declines. Proc Natl Acad Sci USA 2017; 114(30): E6089-96.
[http://dx.doi.org/10.1073/pnas.1704949114] [PMID: 28696295]

[38]   Biggs H. Extinction Crisis 'Poses Existential Threat to Civilization,'. BBC 2020.

[39]   Climate Change: How Do We Know? NASA Global Climate Change.

[40]   Rosen R. The Climate Is Set to Change 'Orders of Magnitude' Faster Than Any Time in the Past 65 Million Years. Atlantic 2013; 2.

[41]   Diffenbaugh NS, Field CB. Changes in ecologically critical terrestrial climate conditions. Science 2013; 341(6145): 486-92.
[http://dx.doi.org/10.1126/science.1237123] [PMID: 23908225]

[42]   Saltre R, Bradshaw CJA. Are We Really in a 6th Mass Extinction? Here's the Science. Science Alert 2019.

[43]   Field C B, *et al.* Climate Change 2014 Impacts, Adaptation, and Vulnerability, Summary for Policymakers. IPCC.

[44]   Environmental Scarcity and the Outbreak of Conflict. Population Reference Bureau 2001.

[45]   Adger WN, *et al.* 2014: Human Security.Climate Change 2014: Impacts, Adaptation, and Vulnerability Part A: Global and Sectoral Aspects Contribution of Working Group II to the Fifth Assessment Report of the Intergovernmental Panel on Climate Change. Cambridge, UK: Cambridge University Press 2014; pp. 755-91.

[46]   Rotman D. Hotter Days Will Drive Global Inequality. MITS Technol Rev 2016; 120(1): 20.

[47]   Global Weirdness. New York: Pantheon 2012; p. 144.

[48]   Rotman D. "Reviews, Hot and Violent," MIT Technology Review, vol. 119, no. 1.

[49]   Gleditsch NP, Nordås R. Conflicting messages? The IPCC on conflict and human security. Polit Geogr 2014; 43: 82-90.
[http://dx.doi.org/10.1016/j.polgeo.2014.08.007]

[50]   Bohannon J. Social science. Study links climate change and violence, battle ensues. Science 2013;

341(6145): 444-5.
[http://dx.doi.org/10.1126/science.341.6145.444] [PMID: 23908197]

[51]   Ibid.

CHAPTER 6

Where Heat Trapping Gases Come From

"What's done cannot be undone."

—Lady Macbeth, Shakespeare

Place in decreasing order the greenhouse gas emissions the following are responsible for:

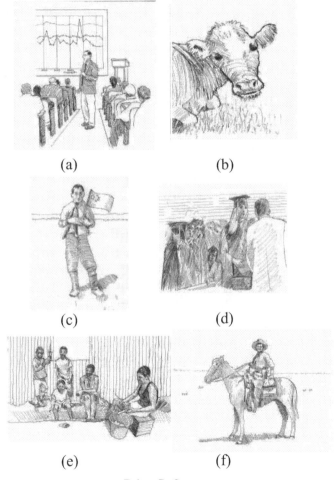

(a)   (b)

(c)   (d)

(e)   (f)

Every year we put into the air an amount of CO_2 equal to the weight of the entire earth's population—sixty-five times over. That's a lot of anything, but when it is CO_2, it is more than we can afford. A little is okay, and in fact a little CO_2 is as necessary for our survival just as too much threatens it. Too little CO_2 and we will have another ice age. But too much will bring on a global warming Armageddon. We need to have enough CO_2 to hold some heat in but not too much or too little. And that's what we have had for over ten thousand years. Earth was in the Goldilocks zone with just the right amount.

We must keep it that way, but over the last one hundred years we have been struggling with an unhealthy increase and the three bears are noticing. We can't change the past, so there is no point worrying about it or pointing fingers. But we can change the future and the first step is to understand the present; what our greenhouse gases are and where they come from.

CHAPTER CONTENTS

It Took a Revolution

You wouldn't think boiling a pot of water for your morning tea would profoundly and irrevocably change the world, now, would you? Get real. Of course it doesn't. How do these ideas ever get into your head?

But it does provide a convenient analogy for what did. In the mid-1700s the English rediscovered that steam could be used for much more than making tea. Since the steam from boiling water expands two thousand times, it can push against objects and move them with extreme force. And if you can move something, you can effortlessly do work without getting up from your breakfast table. It kicked off the Industrial Revolution, which helped create today's amazing technologies, lifestyles, health, and wealth.

The problem though, is that it takes fossil fuels to create all that steam. First wood, then coal, and now natural gas, diesel, kerosene, gasoline, and more coal to keep those factories humming and airplanes flying. As our population increased, enabled by the Industrial Revolution, more fuel was needed and as we moved into cities and traveled, still more fuel was needed and more CO_2 was created. All from a simple cup of tea.

Going on a Carbon Diet

The Intergovernmental Panel on Climate Change (IPCC) reported that we must keep the global warming induced temperature rise below 1.5°C (2.7°F) to avoid serious consequences.[i] And they even prescribed how to do it. We must cut our net greenhouse emissions in half by 2030 and then go to zero by 2050. All good advice, but the only important thing about global warming, and avoiding the 1.5 threshold, is the total amount of greenhouse gases swimming in our air. Once we exceed that limit, it's off the cliff we go. No matter when.

I'll explain with a simple analysis. Before we industrialized and started belching out all this CO_2, there was a fixed amount of it and it never changed (at least not in the past ten thousand years). As nature created more, say from trees dying, other vegetation soaked it up to grow. CO_2 is plants' food and the sun their energy. The balance was unchanging and perfect, a fairy tale almost. Back then the CO_2 concentration was 280 parts per million (ppm) and the air held 2.4 trillion tons of it.

Then along came us clever Homo sapiens and we began burning wood and then coal to drive our steam engines. We built factories, our population exploded, and we became rich. This drove us to pump more CO_2 into the air than nature could remove. We are now at 410 ppm and our air is holding 3.5 trillion tons of CO_2, which is an increase of 1.1 trillion tons over that halcyon preindustrial era.

But to stay under the 1.5 threshold we need to limit the total CO_2 accumulated in our air to 4.6 trillion tons. Which means we have left only another 0.56 trillion tons that we can emit before we cross the 1.5 threshold.

And that is the only thing that matters: the cap of an additional 0.56 trillion tons above which we dare not go. It doesn't matter when we exceed it, be it tomorrow, 2050 or 2100. The key is we must never exceed it.

But we are currently pumping into the air 35 billion tons of CO_2 each year. Putting this together, the following figure shows what's going to happen.

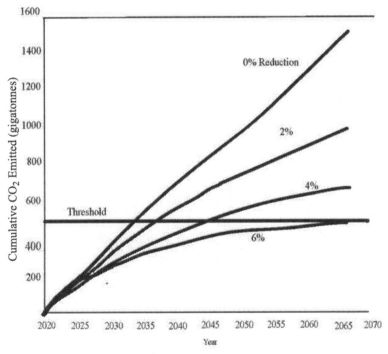

If we do nothing and continue with our current ways, we will reach the 1.5°C limit by 2035, and if we keep on going, we will get to 2°C (3.6°F) by 2050 and by the end of the century 3.5°C (6.3F). Now that's driving off a cliff with rocket assist. If you believe everything I've written to this point, and I think it is close to true, then the path of no action leads us toward ruin.

---

[i] The IPCC is a worldwide nonpartisan organization with thousands of volunteer experts who review the data and peer reviewed published reports on global warming. They periodically release detailed findings, and most governments use this for policy decisions. The temperature rise given is compared to the average measured temperature from 1850 through 1900.

If we get on the ball and reduce our emissions by 6 percent every year, we will stay below the threshold and our bacon is saved. Anything less than a yearly 6 percent reduction doesn't cut it.

Interestingly, at 6 percent we do in fact cut our emissions in half by 2030 as the IPCC wishes, though we go beyond 2050 for a complete reduction.

This also means when we hit the threshold either our greenhouse gases need to zero-out for all time or the excess is removed. At that point we can't digest any more CO_2.

You might ask, "Why do we need to cease all CO_2 emissions after that? Surely eating a small piece of chocolate cake now and again does no harm." The problem with global warming is like a person with a disease that prevents them from burning calories. Even a single cornflake will add weight. So they just keep getting bigger and bigger without end. That's exactly the problem with CO_2. It lasts forever, for all practical purposes. Once a CO_2 molecule is put in the air, it remains there. Put two in and now you have two more than you did before. That is why we cannot add any more past the threshold.

So if we are to keep the global warming wolf from our door, we need to decarbonize, that is, eliminate all fuels containing carbon (or remove the excess CO_2 from the air, which is not so easy but could help). And we must accomplish this before we exceed the remaining 0.56 trillion tons of CO_2.

"But must we decarbonize?" you ask. "Aren't there other ways to achieve the same thing?"

Yes, we must, and no there aren't.

Yoichi Kaya took an interesting and intuitive approach, using a simple equation that demonstrates this point. He showed that there are only four items that will change our CO_2 emission: population, affluence (how rich we live), the carbon content in our fuels, and how efficient we are in burning those fuels [1].

$$Change\ in\ CO_2\ Emissions = \frac{(Population)(Affluence)(Fuel\ Carbon\ Content)}{(Energy\ Efficiency)}$$

If we want our CO_2 emission to go to zero, then one item in the numerator must go to zero.

Population is not going to be one of them. At least one hopes. While the population in developed countries has leveled out, it is rapidly increasing in less developed regions [2], so if anything the world's rising population will increase CO_2 emissions.

Affluence? Come on, let's be real. We are not going back to living in trees. Our standard of living has been increasing across the board, which is a good thing for those benefiting, but not so good for global emissions.

We are surprisingly efficient in burning our fuels, though there are certainly considerable strides to be made. Nonetheless, we are not going to increase our Energy Efficiency so much that it drastically reduces our CO_2 emission. And no matter how well we do, it is not going to eliminate those emissions.

Thus, Fuel Carbon Content is all that remains. We must decarbonize completely and rely only on renewables and nuclear energy production. We need to also capture and store the carbon from any remaining fossil fuel sources. This is the only path available.

"And why 1.5°C?" True, there is nothing special about it and it could have just as easily been 1.4 or 1.6. But 1.5 is close enough that if we stay below it, we will be spared the worst. We will not be unscathed and our planet has already taken significant damage, but to go higher is to greatly increase our risk of civilization changing consequences and the higher we go, the greater the chances of it happening and the worse it will be.

CO_2 is the Greenhouse Gas King

Of the greenhouse gas emitters, CO_2 is the largest at 81 percent, dwarfing the others [3]. So, it would be enlightening to know who or what is most responsible for it. It's the same if someone is pelting us with rocks. It's helpful to know the source.

Of course, CO_2 comes from burning fuels, but who or what burns the most? For this exercise, it is convenient to split the country into four categories:

- Transportation—Planes, trains, and automobiles (ships and other vehicles too).
- Industry—The factories that make all the things we buy and use.
- Residential—Our homes.
- Commercial—Office buildings, shops, malls, government buildings, and the like.

The following tables show the 2020 energy use and the CO_2 emissions for these sectors [4]:

Sector energy use.

Sector	Fossil Fuels (Quadrillion BTU/Yr)	Electric Energy (Quadrillion BTU/Yr)	Total (Quadrillion BTU/Yr)
Transportation	24.2	0.1	24.3
Industry	22.1	9.0	31.1
Residential	6.5	14.3	20.8
Commercial	4.3	12.4	16.7
Total	57.1	35.8	92.9

CO_2 Emissions for Each Major Sector.

Sector	CO_2 from Fossil Fuels (Millions of Tons/yr)	CO_2 from Electric Generation (Millions Tons/yr)	Total CO_2 Emissions (Millions of Tons/yr)
Transportation	1,791	3	1,794
Industry	1,044	399	1,443
Residential	351	636	987
Commercial	250	554	804
Total	3,435	1,593	45,028

Transportation is the biggest emitter. Within transportation, the largest emitters are the cars, SUVs, minivans, and light trucks we drive. No, an SUV doesn't emit more emissions that a 767 jetliner, but because there are so many more vehicles, 284 million (2019), and so few jetliners, 5,882 (2020), what vehicles lack in punch they make up in quantity.

Transportation is not the biggest energy user; that crown goes to industry. But transportation relies almost exclusively on fossil fuels, since there are, relatively speaking, very few electric cars (So far. But the number of electric cars is expected to increase significantly).

And there is some good news here as the average gas mileage of all vehicles has doubled since 1975 and CO_2 emissions have almost been cut in half [5]. Some of that is due to improved technologies and some are the auto industry's drive to reduce the vehicles' weight by replacing steel with aluminum.

There is also some bad news. Unfortunately, we are driving the lower end of the average. The table below shows the fuel economy of different types of cars and we are driving more of the worst ones, the SUVs, minivans, and pickup trucks.

Car-Like and also on Truck-Like. This was my oversight.

Category	Type	MPG
Cars	Sedan/Wagon	31
	Car like SUV	27
Trucks	Truck like SUV	23
	Van/Minivan	23
	Pickup	19

And then there is still more good news since the number of electric vehicles is steadily increasing. But understand, electric vehicles are not emission free. No matter what anyone tells you. They are certainly better than gas and diesel cars, but not zero. Why? Where do they get their electricity that powers them? From electric generating power plants. Where do electric generating power plants get the electricity to send over miles of wire to charge these electric vehicles? Mostly it is from burning fossil fuels, which make up 63 percent of our electric generation.

However, electric vehicles could be emission free whereas gasoline- and diesel-powered cars can never be. The key to emission free vehicles is to decarbonize electric generation. That is, stop burning fuels that contain carbon, which is entirely possible. In fact, looking at the table on CO_2 emissions, if electric generation were completely decarbonized, that is, if they were powered by only renewables and nuclear sources, their CO_2 emissions would go to zero, resulting in a CO_2 reduction of 1,593 tons each year in the U.S. But more on that in Section IV.

"As an Aside, you can distinguish fossil-fired electric generating plants from nuclear ones by the exhaust stacks used. Each will have those distinctive-looking hyperbolic cooling towers but nuclear plants have no need for exhaust stacks because they have no gases to vent. You can tell in the following figure fossil-fired plant from the two tall skinny stacks on the right-hand side of the photo that are venting the exhaust gases from the fossil fuel combustion. The steam seen is from the water vapor (produced from the hydrogen in the fuel) as it hits the cooler outside air and condenses".

Where are the CO_2 emissions heading? While the U.S. emissions are showing a flat and slight declining trend, global emissions have been steadily increasing. (See the following figure [6]) Global CO_2 emissions have continuously climbed from 1990, which was 22.6 gigatons per year and in 2017 it was 36.4 gigatons [7], for an annual increase of 1.8 percent, a troubling trend.

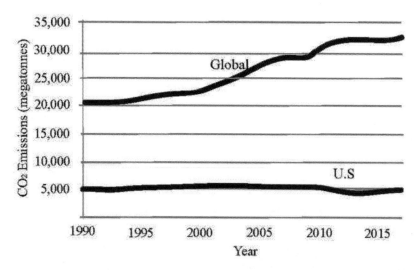

China is the largest emitter by far. They spit out 27 percent of the worldwide greenhouse emissions, more than twice of what the U.S. does, and in fact more than all the developed nations combined. On a per capita basis, they aren't all that bad, but that's because they have so many people, 1.4 billion at last count. And most of them don't live the luxurious, CO_2-spewing lifestyles that most folks in highly industrialized countries do, so their individual contributions are quite small. At least for now.

The following shows China's total CO_2 emissions since 1990 compared to other parts of the world. Up until 2001, their emissions were below North America and Europe, but thereafter it began a steep incline as the economy boomed and they became more industrialized.

Methane

While CO_2 may be the poster child for greenhouse gases, methane is not far behind accounting for 10 percent of the total. One of the two largest sources is methane leaking into the air when handled and transported. Natural gas, which is essentially methane, is lighter than air and is moved about the country in over 2 million miles of transmission pipes. This gives it ample opportunity to leak out, and leak it does, at 100 billion cubic feet per year—the equivalent of the CO_2 emissions from thirty-seven coal-fired power plants [8].

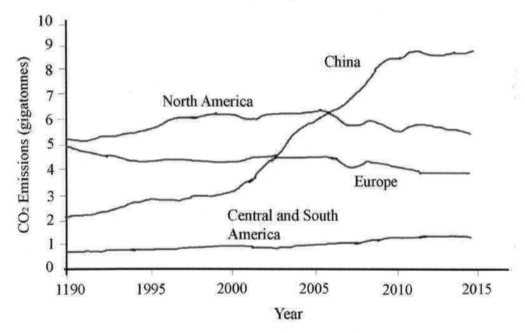

The second major source of methane is agriculture and most of that comes from cow burps from the almost 100 million cows in the U.S. Yes, cows produce and then belch out methane and maybe from their other end too. The next time you eat a burger picture a cow weighing fifteen hundred pounds burping in your face and remember fruits and vegetables don't.

So why does CO_2 get all the attention over methane? A few reasons. For all of its bark, methane isn't as bad as it seems (though it is by no means good). First, methane lasts only about a decade compared to CO_2 being around for centuries. Second, there is much less of it in the air, a couple of ppm as compared to 410 ppm for CO_2.

Third, as greenhouse gases go, methane is the one you want. The leaks are relatively easy to deal with, as it just requires finding and plugging them. It doesn't require any sci-fi like technologies to be developed. Of course, this takes money, so if it there isn't a financial payoff greater than the cost, it might not get done despite the global warming consequences.

As for agriculture, the same thing applies. It is not technically difficult. Just eat less meat and there will be fewer cow burps. Easier said than done, I realize, but at least there are no technical roadblocks.

The Bit Players

There are other greenhouse gases besides CO_2 and methane, but they have less of an effect and are not as important. That is not to say they shouldn't be addressed, but the bulk of our efforts are better spent on the real influencers. Still, they are worth considering and should be understood. Interestingly, some of them are much more potent than either CO_2 or methane, but what makes them bit players is that there is much less of them in the air and they disappear quickly.

A convenient way to evaluate them is through their global warming potential. The following table shows how much more potent these gases are compared to CO_2 over one hundred years (the second column) and how much are in the air compared to CO_2 (third column). The CFCs are more potent than CO_2 by up to twenty-four thousand times, but since they are only one millionth the concentration of CO_2, they have a much smaller true impact. Similarly for nitrous oxide, ozone, and CFCs which are quite dilute and thus not significant contributors.[ii]

Greenhouse Gases Compared to CO_2 over 100 Years.

Gas	Global Warming Potential Compared to CO_2	Approx. Concentration Compared to CO_2
Methane	32	One Millionth
Nitrous Oxide (N_2O)	280	One Thousandth
Chlorofluorocarbons (CFC)	1,000 to 24,000	One Millionth
Ozone	1,000	One Ten Thousandth

Soot, or carbon black, is not a greenhouse gas since it is not a gas at all, but it does contribute to global warming, so it needs to be discussed. Soot consists of small solid particulates formed when fossil fuels are incompletely burned because there are not enough oxygen molecules available to combine with each of the fuel molecules. Hence, some of the unburned fuel turns into solid carbon particles. Black smoke coming off an auto exhaust or from a factory chimney is a visible example. In addition to fuels, forest and brush fires also produce soot.

Soot does its dirty work on the ground when it covers snow and ice. It is black or brownish in color so it readily absorbs the sun's energy the same way black clothing absorbs the sun's energy, making us warmer. The soot then sends some of its newfound heat to the snow and ice beneath it. Further, by covering the snow and ice, it also blocks their excellent reflective cooling, warming them even more. All in all, this pushes the earth's energy balance in the wrong direction, leading to increased global warming.

Soot has a complex effect on clouds when suspended in the air. It can increase the clouds' reflectivity, which is a good thing for us [9], but it also absorbs the sun's incoming energy, which is not. There is uncertainty if soot increases or decreases cloud cover, so the jury is still out on that one.

Last on the list is water vapor which is not considered a greenhouse gas even though it blocks Earth's outgoing radiation. However, in this case it's more the victim than the aggressor since it is the puppet controlled by the CO_2 master and does not independently influence global warming. As the American Chemical Society states, "It does not control the Earth's temperature, but is instead controlled by the temperature. . . . If there had been no increase in . . . CO_2, the amount of water vapor . . . would not have changed [10]".

The important point is that without CO_2 in the first place, water vapor would have no additional effect on global warming. So it is a symptom not a cause.

For The People and by the People

Without people there would be no greenhouse gas increase and no global warming. This is stating the obvious, I know, but the details can be illuminating. The following figure shows that the world's population and the total CO_2 emission from burning fuels closely follow each other over the past 2,000 years [11]. Makes sense. The more people there are, the more food is eaten, more fuel is burned to heat homes, and more gadgets are acquired, all of which take energy.

[ii] CFCs come mostly from the refrigerants used in air conditioners and refrigerators. N_2O, also known as nitrous oxide or laughing gas, comes mostly from fertilizer and animal waste. Ozone comes from gas- and diesel-powered vehicles and from fossil-fired power plants.

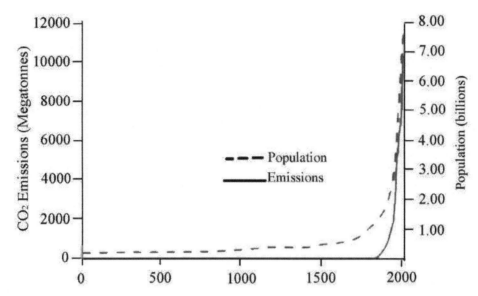

The real jump started right after the Industrial Revolution, somewhere around the mid-1700s. As the world industrialized, it was able to support a larger population which then burned even more fuel.

It took us nineteen hundred years to get to one-half billion people and then only one hundred years after that to add another six billion. Whatever our faults, reproducing is not one of them.

But what really makes the difference is not just the number of people but how high on the hog we live. The richer we are, the more stuff we buy, the more miles we travel, the more food we eat, the bigger are the houses we live in. Rich people produce greenhouse gases like nobody's business. "The best predictor of someone's carbon footprint is their income. American professions, academics, and business people are among the world's greatest carbon emitters [12]". Wealthier peoples' homes produce about 25 percent more greenhouse gases than do their less well-off neighbors, and in some Los Angles neighborhoods it is fifteen times as high [13].

And it goes across liberal and conservative lines, one of the few things that bridge our political divide.

Connecting the Dots

1. In the 1700s the Industrial Revolution profoundly changed the world, leading to the technological wonders we have today. But at a cost. A significant cost. The fossil fuels needed to propel the revolution have led to an enormous release of

greenhouse gases over the past three centuries, leading to irreversible global warming.

2. This now forces us to eliminate all fossil fuels to stabilize the damage already done. And we must accomplish this in the next three decades. It isn't going to be easy.

3. Of all the greenhouse gases, CO_2 is the worst. Mostly it is released from vehicles, planes, ships; most anything that moves and carries people or cargo. But the industry is also a large emitter as are our homes and commercial buildings.

4. Methane, or natural gas, is the next biggest emitter and comes from leaks while being transported in pipes and from agriculture. But methane is easy to fix, at least in the sense that no technological breakthroughs are required.

5. Cow belches also emit methane, so, as the commercial says, "Eat more chicken."

6. Lesser greenhouse gases (and soot) are also eating away at our well-being, but while not unimportant, they are certainly far less important than CO_2.

7. Our insatiable appetite for wealth, and then just a little bit more after that, is driving global warming effects.

Now to answer the quiz from the chapter's beginning:

a. The professor is the largest emitter as are other wealthy individuals.
b. Cows probably come next, what with their rude burping and other nasty emissions.
c. The newly graduated college student is not an important greenhouse gas emitter. Yet. But she will be as she enters the workforce and becomes successful.
d. Our Chinese friend, as an individual, can't compete head to head with the above three. But add all of the Chinese together and you have a juggernaut of greenhouse gas emissions. The most in the world.
e. A poor family from a third world country emits very little greenhouse gases (maybe some from their charcoal cooking stove).
f. The cowboy from the 1800s is clear of all blame. Back then we hadn't yet arrived at the disastrous emissions facing us today.

Afterthought: Cows on Mars?

NASA has measured methane in the Martian atmosphere at concentrations lower than on Earth but significant at 45 parts per billion [14]. Since there is no industry on Mars, where does it come from? Uh oh, I hope I didn't just spark a conspiracy theory that NASA is hiding a thriving Martian civilization. Anyway, on Earth its cow burps that account for much of our airborne methane, so why wouldn't that be the same on our nearby neighbor? Imagine, cows roaming the Martian landscape. Now that's a conspiracy theory I can get behind. Okay, so maybe not cows but microbes also produce methane and the measurements are seasonal which mirrors life's tendencies. Unfortunately, the explanation may be more mundane and it could come from tiny meteors striking the planet. Still, I'm rooting for the cows.

Reference

[1]   Kaya Identity Wikipedia Available from: https://en.wikipedia.org/wiki/Kaya_identity/

[2]   "World Population Prospects, 2019," United Nations, Department of Economic and Social Affairs Population Dynamics.

[3]   "Greenhouse Gas Emissions," US Environmental Protection Agency.

[4]   Energy and the Environment Explained EIA Available from: https://www.eia.gov/energyexplained/energy-and-the-environment/where-greenhouse-gases-come-from.php/

[5]   Automotive Trends Report. EPA.

[6]   Data and Statistics. The International Energy Agency (IEA).

[7]   Chestney N. Global Carbon Emissions Hit Record High in 2018: EIA Reuters 2018.

[8]   Schwartz J. Methane Leads in Natural Gas Supply Chain Far Exceeds Estimates, Study. New York Times 2015.

[9]   Kerr RA. Global warming. Soot is warming the world even more than thought. Science 2013; 339(6118): 382.
        [http://dx.doi.org/10.1126/science.339.6118.382] [PMID: 23349261]

[10]   Sheperd M. "Water Vapor *vs.* Carbon Dioxide: Which 'Wins' in Climate Warming?" quoting the American Chemical Society's ACS Climate Toolkit. Forbes 2016; 20.

[11]   Population data and emission data taken respectively from https://www.worldometers.info/world-population/world-population-by-year/ and https://cdiac.ess-dive.lbl.gov/trends/emis/tre_glob_2013.html/

[12]   Lindberg E. By Erik Lindberg: Six myths about climate change that liberals rarely question. 2016.

[13]   "Gas Houses," Scientific American, November 2020.

[14]   Hand E. Mars Methane Rises and Falls with the Seasons. Science 2018; 359(6371).
        [http://dx.doi.org/10.1126/science.359.6371.16]

Part III. Why We Believe Global Warming Is Real and Significant

While part II showed why global warming is occurring, part III covers why we are so sure of it.

CHAPTER 7

Global Warming is Undeniable. Here's Why

"It's really not rocket science. You look at the evidence."

—*Michael Sherwin, Acting U.S. Attorney for the District of Columbia*

Nero Wolfe, Philip Marlowe, Jane Marple, Hercule Poirot—all understood a basic truth: evidence convicts the murderer. Nothing else can.

The same approach, the only possible approach, tells us that global warming is real and we are the cause. It is based on sound scientific principles, including satellite data, modern recorded measurements, complex models, prehistory data, and thousands of experts devoting their professional lives studying it. The evidence and facts are overwhelming and come from many independent and converging sources. There are no conspiracy theories, propaganda or flashy social media stunts, or a run at our emotions at play here. Just the truth.

And while there are no absolute truths in this imperfect world of ours, there are enough facts and evidence to lay claim to global warming's reality. Not absolute, but as close as mortals can get.

CHAPTER CONTENTS

How Far Can Tom Brady Throw a Football On The Moon?

Now there's an interesting question. Any guesses? The direct approach is to rocket Brady to the moon and have him throw a football and measure its distance. Easy peasy, you might say, so let's do it. After all, who wouldn't want to know how far a football will go on the moon. But there is a logistical hitch to dispense with first. The spacesuit needed on the moon is too confining for throwing a football, so it will have to come off, leaving Brady only 15 seconds to live. Not a problem; he can wear a quick breakaway suit, and since his release is 2.5 seconds, we can get a good six data points before he ascends to the great gridiron in the sky.[i]

I know Brady will go all in since he will then be able to throw a football farther than Aaron Rogers ever could, and it will add to his already sizeable fifty-four NFL records. He'd gladly use his last fifteen seconds for that. After all, he's not called the GOAT for nothing. Though I think the monkey wrench in all of this will be Gisele; it'd be just like her to object. And his agent too. But I bet Bill Belichick would give a thumbs up. Now, there's a guy who sees the big picture.

Okay, forget strapping Brady to an Atlas rocket. There's another way. We can find the answer by building a mathematical model that includes the physics of Brady throwing a football. We can model gravity pulling down on the football, air resistance (on Earth) slowing it down, his release angle and velocity, the weight and size of the football, and the forces acting on the football.

Which is what I did. The following figure shows the football trajectory after his release on the moon and, for comparison, in Tampa Bay. He will throw the football about 500 yards, with a hang time of twenty-four seconds, reaching a height of 127 yards. Take that, Aaron Rogers.

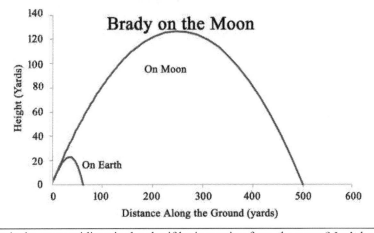

---

[i] Which direction is the great gridiron in the sky if he is starting from the moon? Isn't he already in the sky?

But Brady will have good days and not so good ones, so it would be better to bracket his performance by looking at how things might go wrong or be above average. For instance, I varied his throwing angle and release velocity by 10 percent up and down, with the results in the figure. Now his performance will be as low as 400 yards and as high as 600 yards. Further, I could run this simulation thousands of times, varying slightly all of the conditions, and come away with an excellent idea of his performance.

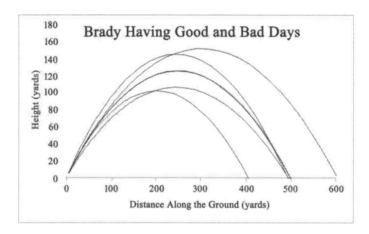

Do you believe my results? You shouldn't. At least not until they are verified. All mathematical models must be verified to provide the confidence they are reasonably accurate; otherwise, they are useless. We cannot base decisions on models we don't trust.

No need to go to the moon to verify the model since it will predict the distance of a thrown football on Earth in many situations that we can measure directly and decide if the model works or not. For instance, we could take Brady to the top of Mount Everest, where gravity and air resistance are just a bit lower than at sea level, and measure the distance and compare it to the model. We could have athletes of different abilities, such as high school and college players, throw a football under different conditions and see if the model predicts their performance.

We could deflate the football, which will reduce its weight and area, and compare what Brady actually throws to what the model predicts. In fact, we already have the test data since the football was deflated when Brady played the Colts in 2014. Well, that's a sore subject so maybe we shouldn't go there. Best to let sleeping quarterbacks lie. But you get the picture. By comparing the model's predictions to what is measured, we can nail whether or not it works, and if it is off, what changes might be needed to set it right.

The model can be refined, as time and money allow, programming in other influences on Brady's performance, such as air humidity and wind in Tampa Bay, the moon's rotation under the football, terrain differences, and so on.

What have we accomplished? We had a question that could not be answered by direct measurement, so we built a mathematical model simulating it using accepted scientific principles. We verified the model by comparing its results to situations we could easily measure. We produced an answer to our question as well as how sensitive it is to changing conditions. We discovered that Brady would most likely throw a football 500 yards on the moon, but it might also be between 400 and 600, though less likely. We identified areas of the model that could be improved. And we are reasonably confident in our answer.

Which brings me to the point of this section. This is exactly how climate models work in predicting global warming effects due to CO_2 emissions. Climatologists build models the same way I did, though with far more complexity. Where my Brady model was a simple Excel spreadsheet, these models contain millions of lines of code. They include the same fundamental laws of physics but also the dynamics of the atmosphere, including the effect of CO_2, the effect of clouds, the earth's energy balance, melting glaciers, land, ocean and ice features, and more. The following figure shows some of what is included in the models.[ii]

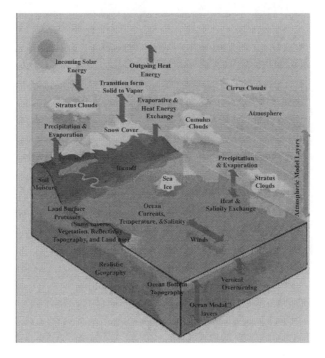

[ii] Figure is attributed to © 2022 UCAR.

Why are models needed in the first place? Because some things cannot be directly measured, especially when it is as complex as the climate. I can't take Brady to the moon. And I can't travel into the future to see what the climate will be. But models can.

Just as I explored Brady's performance by varying his release angle and the football's initial speed, climate models perform thousands of simulations, changing each of the far more numerous variables to arrive at an average projection, how sensitive the model is to each variable, and how the results might change as people change their habits.

The climate models have been verified in a fashion similar to what I proposed for Brady. One powerful example is given by the IPCC in which they calculate the temperature rise from 1900 to 2000 with and without CO_2. They then compare each result with the actual measured temperature rise [1]. The only way the model accounts for all the global warming effects is to include the CO_2 we have been creating. Pull that out of the model, and it can't duplicate anything. Put it back in, and viola, the models match the real world quite nicely.

Similar results shown in the following figure come from NASA [2]. The model using CO_2 matches closely the actual temperature measurements. When the CO_2 was removed (not shown in the figure), the model presented a flat, unchanging temperature throughout the period shown.

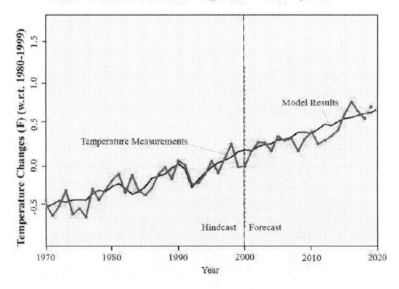

An important verification of climate models is that they accurately predict the past climate. The example above goes back to the 1900s, but the models predict the climate back hundreds of thousands of years.

Climate models can accurately simulate past and present climates, are built by unbiased researchers, and there are many independent models that agree among themselves. They are an extraordinarily valuable tool to guide policy and our understanding of global warming. And they will get better as the models are refined, more data becomes available to test them with, and computers become more powerful.

Who are the model meisters? A list of U.S. modeling efforts, as of 2013, is given in the following table [3]. A more complete list of worldwide climate modelers can be found in the paper "List of Global Climate Models [4]".

U.S. Climate Modeling.

Where	Funding	Notes
Community Earth System Model/NCAR	NSF-DOE	Most Extensive Effort
Geophysical Fluid Dynamic Laboratory	NOAA	Major Model and includes Princeton University
National Center for Environmental Prediction	NOAA	Short- and Medium-Term Predictions
Goddard Institute for Space Studies	NASA	Global and Regional Climate Sensitivity
Goddard Earth Observing System	NASA	Model with Space Observations
Proposed DOE Model	DOE	Would Focus on High Resolution

Are the models 100 percent accurate? A show of hands. Glad to see most readers say no. Of course not. Nothing in life can ever be perfectly accurate, and certainly nothing in science. Are they accurate enough for us to see the future with confidence and know what actions are needed? Now the answer is a resounding yes.

Climate models—don't do global warming without it.

"to Measure is to Know"—lord Kelvin

The best arbiter of global warming is direct temperature measurements, and these have been available to some degree, at least for the past 350 years. The thermometer was invented in 1660, resulting in an unbroken temperature record in

England since then. Weather stations were established in the U.S. and in England in the 1800s. From the 1850s onward began the modern temperature record era, and by 1957 worldwide coverage was consistent and then became further improved after 1980 when satellite measurements came on line [5]. There are now thousands of land stations continuously recording temperatures and thousands more on ships, plus the more recently added satellite measurements.

The key is figuring out how to handle this avalanche of data. Statistics and averages come to the rescue. All of these temperature measurements are averaged across the globe and then averaged again over a single year. From NASA, there are thirty-two thousand measurement locations [6]; assuming each record a temperature every minute, then there will be 17 billion measurements every year. Averaging all of their results in a single temperature of the entire planet for an entire year.

This is critical to understand, as it points out that climate temperature measurements are neither local nor immediate, as you would find listening to your neighborhood weather channel. Weather is a specific event at a specific location at a specific time. Say, the temperature is at 6 PM in your city. Climate is an average of each of these specific weather events across the entire globe for an entire year.

This is why an unusually cold or hot day says nothing about global warming. It just means the weather for that day was unusually cold or hot, nothing more. Rather, it is the global average over a period of time compared to an earlier time that tells the story.

This is shown in the following figure, which shows the annual worldwide temperature change from 1880 to 2020 [7]. Somewhere around the early 1900s, temperatures rose and, after a brief lull in the 1950s, continued to rise. The worldwide average yearly temperature was 13.5°C (56.3°F) in 1850 and rose to 14.5°C (85.1°F) in 2003 for a 1°C (1.8°F) total rise.

Remember, the 1.8°F rise is across the globe, so it represents an enormous amount of additional heat added to our planet since 1850. To put it in perspective, a one-degree drop was all that was needed to push Earth into the Little Ice Age, and a five-degree drop was enough to bury most of North America under a massive tomb of ice twenty thousand years ago [8].

Annual Worldwide Temperature Rise

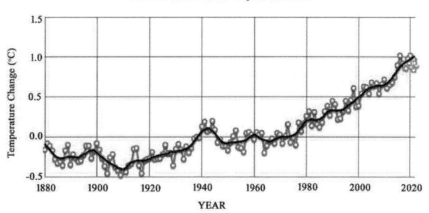

Satellites are the new kid on the block, relatively speaking, and they come crammed with instruments. Looking down from their perch several hundred miles high, they can see the entire planet and measure heat entering and leaving the earth, the change in ice coverage, CO_2 content in the atmosphere, temperature, and more.

A partial list from Wikipedia shows there are sixty-one active satellites orbiting the earth, measuring aspects of climate and weather [9].

One example—twenty-five years of satellite observations of Antarctica's ice shelves have definitively shown an ice loss of 4 trillion tons since 1994, enough water to fill the Grand Canyon [10].

Another satellite, from the European Space Agency, is FORUM, which stands for earth's Far Infrared Outgoing Radiation Understanding and Monitoring. Clearly they tried to force a name to fit its predetermined acronym, which went less than well. But hey, we shouldn't expect clever prose and good science at the same time. It is planned to launch in 2025 or 2026. Its purpose is to measure Earth's radiation budget which is the sun's short wavelength radiation entering the earth and the longer radiation wavelength outwardly emitted into space by our planet. And how our activity affects each [11]. That'll be good science, indeed.

One more example is NASA's Atmospheric Infra-Red Sounder satellite (AIRS—another forced acronym), which measured Earth's surface temperatures from 2003 to 2017. NASA used this data to confirm the land and sea station measurements were accurate, adding another harbinger of global warming.

Dead Men Do Tell Tales

Imagine going back thousands of years and talking to people long dead about the weather at their campsite. What the temperature was, how much CO_2 was in the air, and how much ice was around them. Sounds as cartoonish as Mr. Peabody's Wayback Machine, but we can do exactly that. In fact, we can go back as far as two million years, long before modern humans even appeared [12].

Ice Cores

One approach has to do with the way buried ice stores its climate secrets. In Antarctica, Greenland, and some high mountain glaciers, snow falls every year and never melts because the temperature rarely climbs above freezing [13]. So each year, a new layer of snow covers and compresses the previous one, and this has been going on for millennia. If you dig down a few inches, you would hit last year's snow; Tens of feet maybe last century's snow, and a couple of miles, snow that was laid down a million years ago.

As the snow turns to ice, it creates air bubbles containing gases present at that time, including CO_2, methane pollen, volcanic ash, and more. These ancient air bubbles almost never escape and remain well preserved. Importantly, the age of the bubbles can be accurately dated using radioactive isotopes, a common and well-developed tool borrowed from archeology. And their temperature at the time they formed can be determined by comparing the different types of hydrogen and oxygen atoms present. Thus, ice cores provide a continuous uninterrupted reading of the atmospheric gases and their temperature, time-stamped from the present to about eight hundred thousand years ago and perhaps as far back as two million years. And they can be correlated to other temperature data recovered by other techniques. The two figures below show ice cores being removed and stored [14].

One important fact discovered from the ice cores is that the amount of CO_2, while variable, never exceeded 300 ppm over the past eight hundred thousand years until modern times, when it spiked to over 400 ppm figure [15]. The spike occurred after the Industrial Revolution but mostly in the past fifty years or so. From other evidence, the CO_2 spike is consistent with the amount of CO_2 emitted from the fossil fuels we have burned.

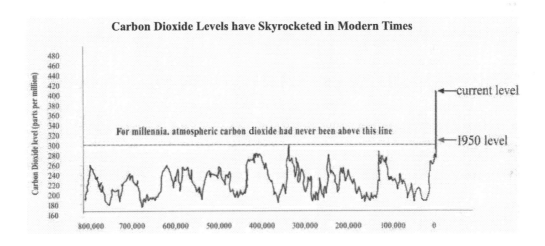

Another important fact is that as the amount of CO_2 increased, so did the temperature, and as CO_2 decreased, the temperature also decreased, providing evidence they are directly linked [16].

Tree Rings

Tree rings provide climate information going back maybe fifty thousand years [17], although ten- to twenty thousand years are more likely. And while tree rings don't contain as much information as ice cores, they are easier to acquire and more numerous and often are part of an archaeological dig or ancient ship. Trees rings are accurate timepieces, as each ring is one year of growth, and their thickness speaks to the climate conditions of that year, including the temperature and amount of rainfall.

Boreholes

A surprising fact is that heat from the surface slowly travels down through the ground without much loss because dirt is an excellent insulator. Since heat is an indicator of temperature, the temperature of the surface also slowly works its way downward along with it and at the same speed. Thus if a thermometer is sent into a hole, it will eventually catch up with the heat that was generated from the surface some time ago. For instance, digging 492 feet will indicate what the surface temperature was 500 years ago and 1,640 feet what it was 1,000 years ago [18]. Despite the excellent insulation, some heat is inevitably lost, so there is always some error in the time the surface temperature occurs. Still, researchers have concluded from this data, coupled with models, that the temperature increase in the last half of the twentieth century was greater than in the preceding four hundred to one thousand years [19].

Other Esoteric Ways

There are many other clever ways to infer the past climate. For instance:

- The size and depth of plant stomata, the pores plants use for gas exchange, change depending on the amount of CO_2 in the air; the more CO_2, the fewer pores needed and the smaller they will be.
- The abundance of different animal fossils in lake and ocean sediments gives hints about past climates, as different animals thrive in different climate conditions.
- The isotope ratios in snow, corals, and stalactites provide reasonably accurate temperature measurements.
- Ancient pollen concentrations tell us about the type and abundance of trees, which indicate the local climate and temperature at that time.

Putting it all Together

Putting all of this together yields the following graph [20]. Over the past two thousand years, Earth's temperature has fluctuated, as seen from the Medieval Warm Period and the Little Ice Age, but nothing compared to the missile-like rise over the past fifty years figure.

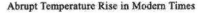

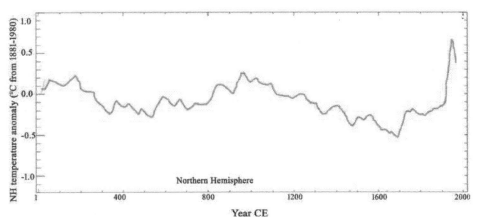

"*Here's an interesting Aside. Since greenhouse gases reduce the heat leaving the earth but have no influence on the sun's incoming radiation, it is expected that nights would be more affected than days. During the day, both the sun and the earth ply their trade. But at night, only the earth is active, whereas the sun doesn't stay up past sundown. And that is what is found: the number of warm nights is increasing faster than the number of warm days [21]*".

Global warming is increasing extreme weather, such as large storms and flooding, creating a new scientific discipline that assesses if global warming is responsible for any of these events. Known as *extreme weather attribution*, there is now enough data to draw conclusions. Researchers using decades of accurate data and running thousands of model simulations believe that under some circumstances, global warming is responsible for some types of extreme weather, such as hot and cold spells, drought, and excess precipitation [22].

For instance, the World Weather Attribution initiative used computer simulations and found that the northern Europe hot spell during the summer of 2018 was twice as likely to have been caused, at least in part, by global warming [23].

I'm a bit skeptical of single-event attributions though I will keep an open mind, especially since it promises splendid entertainment once lawyers get a whiff of it. Saying with some certainty that a calamity was caused by burning fossil fuels is blood in the water and will cause a feeding frenzy among our one million-plus litigious lawyers.

Will energy companies be liable? Who knows. But they have deep pockets, which is every lawyer's fantasy. As F. Lee Bailey said, "The guilty never escape unscathed. My fees are sufficient punishment for anyone." Let the show begin.

Global warming is, unfortunately, alive and well. Take the modern measurements since 1850, add to that the proxy measurements going back more than two million years, stir in the considerable theoretical understanding about the physics and chemistry of it and we have not just the smoke but the fire itself. Global warming is here. And it's here to stay.

Yes, Global Warming has its Uncertainties

Sadly lacking omniscient abilities, researchers have run into some difficult questions that cannot currently be answered but will affect global warming's future (and ours too, I'm afraid). The first of these is the effect increasing temperatures and humidity will have on clouds and, in turn, the effect clouds will have on temperatures and humidity. A bit of circular reasoning and nobody likes those.

Here's the thing. Clouds change the heat flow from both the sun's incoming radiation and the outbound cooling leaving the earth. Dense, low clouds reflect the sun's heat back into space (lessening global warming) and high thin clouds stop heat from leaving the earth (intensifying global warming). The amount of water vapor will certainly increase with increasing temperatures, but which type of cloud wins? Water droplets, which favor the lower atmospheric clouds, reflect more solar radiation back into space than do ice crystals, which favor the upper atmospheric clouds. Models may not be correctly calculating the ice water balance and any flaw could be serious if models are overstating clouds' cooling [24]. Recent observations show that high clouds are winning the race [25], and that's not a good thing at all.

Permafrost, that bugaboo in the frozen wastelands that has locked up tons of carbon, is another one. As the permafrost melts, does it release its imprisoned carbon in a giant burp or in small piddles, or nothing at all? It could make for a very bad day if all that carbon is released in a short time but there is no consensus on how it will go.

Plants' response to global warming is another unknown. Foliage soaks up CO_2, and the more it does, the bigger it gets, and therefore the more CO_2 it needs to sustain its growth. Does more CO_2 increase plant growth so that it absorbs more or do they become bloated and unable to absorb any more the way we are after a Thanksgiving dinner. Plus, global warming is already forcing trees to move farther north which will diminish their appetite for CO_2.

Similarly, oceans are the biggest eater of CO_2, but will they saturate as well? Then what happens if we continue to emit CO_2 but the oceans and foliage walk away from the dinner table?

But the wildest unknown of them all is people. You've seen them; those bipedal anthropoids with huge brain cases capable of sending Tom Brady to the moon but sometimes unable to punch their way out of a brown paper deception. And people will ultimately decide our fate irrespective of the other unknowns. What we do is the key. How much CO_2 will we emit, how many trees will we cut down, and how many more people will we put on this already overcrowded world? We drive everything. How will we behave? No one knows. No one can know.

IPCC: The Go-to Guy

The Intergovernmental Panel on Climate Change (IPCC) was created in 1988 by the United Nations. Its mission is to provide objective, independent information on global warming and its impacts. Thousands of volunteer experts work for it by tracking peer-reviewed literature and other sources for accurate information on global warming. The IPCC is recognized as an accepted authority on global warming, producing periodic reports vetted by numerous scientists and delegates from more than 120 governments [26]. They published comprehensive reports in 1990, 1995, 2001, 2007, 2014, and 2022.

While they don't conduct research themselves, they provide an extraordinary service by aggregating the vast research on global warming, assessing its reliability, providing future projections on where it is going, and recommending limits and actions we should be taking.

In my view, they do the impossible by staying objective and nonpolitical amid a horde of competing geopolitical interests from their 120 members. If you ever doubt that global warming is solvable, just look at what the IPCC has already accomplished, which should give us hope that any impossible task is possible.

Their reports are a bit of a heavy read, though, partly because they need line-by-line consensus from many factions. Still, if you want detailed, unbiased, accurate information on global warming and are willing to wade through it, this is not only the best but maybe the only game in town.

Interestingly, the U.S. was the main driver in its formation during the Reagan administration.

Their latest reports conclude:

1. Global warming has been confirmed since the 1950s, and many of the changes are unprecedented over decades to millennia.
2. Global warming will likely lead to severe, pervasive, and irreversible impacts.

3. Greenhouse gases have increased to levels not seen in at least 800,000 years.
4. It is certain that we are responsible for it (95 percent to 100 percent probable).

For their efforts, the IPCC won the 2007 Nobel Peace Prize "for their efforts to build up and disseminate greater knowledge about man-made climate change and to lay the foundations for the measures that are needed to counteract such change."

Connecting the Dots

1. Climate models are used to predict the effects of greenhouse gases, and they are quite accurate. They are built using accepted and sound scientific principles, they are exhaustively tested under known conditions, and they are peer-reviewed by armies of researchers, there are many of them, and they all agree.
2. But there is no substitute for direct measurements, and of those, we have plenty. Temperature measurements have been available for 350 years, weather stations for 300 years, and, more recently, satellites tirelessly looking down and watching.
3. Plus, we have tools to travel back in time, thousands of years, and sometimes millions of years, to measure the weather back then. This comes from ice cores, tree rings, boreholes, and other ways.
4. Sure, there is some uncertainty. Life is an uncertain gamble and always will be. The same is true for science. But we know enough that even with the uncertainty, we are sure how global warming is occurring and where it is headed.
5. All the data, calculations, researchers' countless hours, and so much more, point to the same thing. Global warming is happening, it is caused by greenhouse gases, mostly CO_2, which comes from fuels we burn, and we are seriously running out of time.
6. The IPCC is the place to go for unbiased, accurate information and is our most trusted source.
7. Oh, and did I mention we are running out of time?

Afterthought: Facts Need Not Apply

Global warming facts are undeniable, but when has that ever stopped people in power from throwing shameless lies at us? Lies are so absurd they will make your head spin as if possessed by a demon. Sure, there are some honest mistakes too, but the tally of Exorcist-like droolings slung at us daily is overwhelming. Here are a few from both categories.

"With all of the hysteria, all of the fear, all of the phony science, could it be that

man-made global warming is the greatest hoax ever perpetrated on the American people?"

—Senator James Inhofe, "The Science of Climate Change" [27]

Will the real hoax please stand up?

"[Global warming denial] has come about due to deficits in facts and reasoned thinking—and can therefore be sufficiently tackled with education."[28]

—Attributed to Steven Pinker of Harvard University and Hans Rosling

So if we are not from Harvard, we must be deficient thinkers. This is as false as the previous quote. It has nothing to do with reasoning or facts, but about beliefs and money. Tone deaf and naïve.

"How bad will climate change be? Not very. Human greenhouse emissions will warm the planet, raise the seas and derange the weather, and the resulting heat, flood and drought will be cataclysmic. Cataclysmic—but not apocalyptic."[29]

—Will Boisvert, "The Conquest of Climate"

Damn that's reassuring. So it won't be the end of humanity, just most of us.

"The climate change argument is absolute crap."[30]

—Tony Abbott, former prime minister of Australia

Do you know who invented the toilet? Thomas Crapper.[iii] No shit. He really did. However, I am now wondering if maybe Tony Abbott should claim the toilet crown.

"Joe Biden said 150 Million Americans died from guns and another 120 million from covid-19."[31]

—Facebook post reported by Nusaiba Mizan in *USA Today*

That's a total of 270 million people who died out of a U.S. population of 328 million. Look to your left and to your right. If the people you see are not dead, than you must be. Nice trick reading this book, Mr. Deadperson.

[iii] Thank you, *Uncle John's Bathroom Reader*, for this fact.

References

[1] Climate Change 2007, the Physical Science Basis 2007.

[2] Buis A. Study Confirms Climate Models Are Future Warming Projections Right NASA 2020. Available from: https://climate.nasa.gov/news/2943/study-confirms-climate-models-are-getting--uture-warming-projections-right/

[3] Science 2013; 341: 13.

[4] List of Global Climate Models. Climate Change in Australia Available from: https://www.climatechangeinaustralia.gov.au/en/climate-projections/about/modelling-choices-and-methodology/list-models/

[5] Solomon S, *et al.* Climate Change 2007: The Physical Science Basis. Intergovernmental Panel on Climate Change

[6] Buis A. The Raw Truth on Global Temperature Measurements NASA 2021. Available from: https://climate.nasa.gov/ask-nasa-climate/3071/the-raw-truth-on-global-temperature-records/

[7] Global Land-Ocean Temperature Index NASA Global Climate Science Available from: https://climate.nasa.gov/vital-signs/global-temperature/

[8] Earth Observatory. NASA Available from: https://earthobservatory.nasa.gov/world-o--change/decadaltemp.php/

[9] "List of Earth Observation Satellites," Wikipedia.

[10] Amos J. "Climate Change: Satellites Record History of Antarctic Melting," BBC, https://www.bbc.com/news/science-environment-53725288/, 10 August 2020; "IceSat: Space Will Get Unprecedented View of Earth's Ice," BBC News, https://www.bbc.com/news/science-environmen--45523524/, 15 September 2018.

[11] Far Infrared Outgoing Radiation Understanding and Monitoring. Wikipedia Available from: https://en.wikipedia.org/wiki/Far-infrared_Outgoing_Radiation_Understanding_and_Monitoring/

[12] Kelly M. Two-Million-Year-Old Ice Cores Provide First Direct Observations of an Ancient Climate. Princeton University 2019. Available from: https://www.princeton.edu/news/2019/11/21/two-millio--year-old-ice-cores-provide-first-direct-observations-ancient-climate/

[13] Ice Cores Wikipedia Available from: http://en.wikipedia.org/wiki/Ice_core/

[14] Core Questions: An Introduction to Ice Cores. NASA Global Climate Change Available from: https://climate.nasa.gov/news/2616/core-questions-an-introduction-to-ice-cores/

[15] Climate Change: How Do We Know? NASA Global Climate Change Available from: https://climate.nasa.gov/evidence/

[16] Kelly M. Two-Million-Year-Old Ice Cores Provide First Direct Observations of an Ancient Climate Princeton University 2019. Available from: https://www.princeton.edu/news/2019/11/21/two-millio--year-old-ice-cores-provide-first-direct-observations-ancient-climate/

[17] Henson R. Climate Change. Rough Guides 2006; p. 192.

[18] Proxy (Climate) Wikipedia Available from: https://en.wikipedia.org/wiki/Proxy_(climate)/

[19] Borehole Temperatures Confirm Global Warming CNN 2000. Available from: https://www.cnn.com/2000/NATURE/02/17/boreholes.enn/

[20] Paleoclimatic Data for the Past 2,000 Years. NOAA Available from: https://www.ncei.noaa.gov/sites/default/files/2021-11/9%20Paleoclimatic%20Data%20for%20the%20Last%202,000%20Years%20and%20Before%202,000%20Years%20Ago%20-%20Oct%202021.pdf/

[21] Cook J. The Scientific Guide to Global Warming Skeptism.

[22]   Cornwall W. Efforts to link climate change to severe weather gain ground. Science 2016; 351(6279): 1249-50.
[http://dx.doi.org/10.1126/science.351.6279.1249] [PMID: 26989227]

[23]   Rainy J. Global Warming Can Make Extreme Weather Worse. Now Scientists Can Say by How Much. NBC News 2018. Available from: https://www.nbcnews.com/news/us-news/global-warming--an-make-extreme-weather-worse-now-scientists-can-n901751/

[24]   Schwartz J. Climate Models May Overstate Clouds' Cooling Power, Research Says. New York Times 2016. Available from: https://www.nytimes.com/2016/04/08/science/climate-models-may-overst-te-clouds-cooling-power-research-says.html/

[25]   Hood M. New Research Models Suggest Paris Goals May Be Out of Reach Phys Org 2020. Available from: https://phys.org/news/2020-01-climate-paris-goals.html

[26]   "Intergovernmental Panel on Climate Change," Wikipedia.

[27]   Available from: https://en.wikipedia.org/wiki/Global_warming_conspiracy_theory/

[28]   Klintman M. Conspiracy Theories: How Belief Is Rooted in Evolution—Not Ignorance. Available from: https://theconversation.com/conspiracy-theories-how-belief-is-rooted-in-evolu-ion-not-ignorance-128803/

[29]   Horgan J. Should We Chill Out about Global Warming? Sci Am 2018; 8.

[30]   Stuff Deniers Say: Six of the Craziest Claims about Climate Change. Climate Reality Report 2015. Available   from: https://www.climaterealityproject.org/blog/stuff-deniers-say-six-craziest-c-aims-about-climate-change

[31]   Mizan N. Fact Check: Yes, Biden Botched Stats on Covid and Gun Deaths USA Today 2020. Available   from: https://www.usatoday.com/story/news/factcheck/2020/07/18/fact-check-joe-bi-en-botched-stats-covid-gun-deaths/5461700002/

CHAPTER 8

Science doesn't Work the Way You Think it Should

A new scientific truth does not triumph by convincing its opponents . . . but rather because its opponents eventually die.

—*Max Planck (1858–1947)*

How would most people define science? From dictionary.com:

"Science is a branch of knowledge or study dealing with a body of facts or truths systematically arranged and showing the operation of general laws."

Which is no better than having a priest describe married life. It doesn't just miss the essence of science; it murders it with a brain-dead, bloodless definition that could take the buzz off a Grateful Dead concert.

Science is a human activity with all of its triumphs, failures, deceit, and heroism. It is as deplorable as the syphilis experiments carried out on black sharecroppers and as glorious as the Apollo space program. It is both noble and tragic, used to cure diseases and design gas chambers. And it is no more systematic than people are robots.

Certainly it deals with general laws and knowledge, but to say this defines science is to describe the Mona Lisa as consisting of paint.

Science is about people. Talented people, to be sure, but with their own Metabeliefs, dreams, quirks, and flaws that find their way into the science they do. This is not a judgment. It is simply the truth. It is how science gets done.

And it is Dr. Jekyll and Mr. Hydian in nature. On the one Dr. Jekyll hand, it is a well-defined and proven approach that has yielded advances and riches for us. On the other hand, Mr. Hyde is messy with warts, errors, limitations, and continual reversals. To understand science is to understand both.

CHAPTER CONTENTS

The Minimalist'S Guide To Science

Science has specific traits that set it apart from non-science; that was developed over many years, starting at least with Aristotle and hitting its stride in seventeenth- and eighteenth-century Europe. Galileo and his contemporaries chaperoned modern science pretty much the way it is today. Sure, we have better tools and more facts to work with, but the way science thinkers think about science hasn't changed since then.

Science works incrementally, building one layer at a time from past successful work. And while it may occasionally make great leaps, Einstein's theory of relativity or Newton's gravity, for instance, is still based on what came before. As Newton said, "If I have seen further than others, it is by standing on the shoulders of giants."

But first and foremost, science is about evidence and, more specifically, that which can be measured. In other words, hard, accurate, uncompromising data. Data is used to either support or falsify someone's idea on how some aspect of the universe operates. Reality is the final judge and executioner of any scientific idea.

Science certainly contains theories, analysis, computer simulations, and the like, but its one and only true master is data. All of the science lives and falls by it. If you cannot measure it, it's not science. It might be proto-science, philosophy, religion, opinion, politics, or even interesting, but not science.

This is not a trivial hair-splitting distinction. It is crucial to understand what science is, what it can do, what it can't do, and where we can be misled into believing lies dressed up in science's clothing. As sham science quickly descends into quackery and drinking bleach will cure autism and the earth is a flat pancake. Science banishes these. Trickery amplifies them precisely because it purposely avoids data that can validate or, more likely, falsify its claims.

Astrology is an example of a pastime that has both science and not even close to science aspects. It uses the planets' and stars' locations (science) to predict if we are finally going to meet that tall dark stranger. But where is the explanation of how the planets' alignment will predict our lives, and where is the data to support it? Neither exists, of course. It must be rejected as science and, therefore, as a truthful description of reality since no data exists to confirm it, yet much data exists to refute it.

A good comparison of science versus not science is evolution versus intelligent design. There is nothing wrong with either, but one is science and the other is not. Evolution is based on countless confirming facts, including fossil finds that show a clear and consistent picture of evolution and, more recently, DNA evidence to further support it.

By contrast, intelligent design has no facts to verify it, so it does not pass the test of being falsifiable. It is not science. Intelligent design is a belief belonging quite comfortably in religion's camp. That is its home and a valid one for it. Crossing the railroad tracks into science's neighborhood is a fossil out of water.

"As an Aside, a fascinating example of science is Michelson and Morley's light-speed experiment. Using a clever setup, they measured the speed of light to be about 186,000 miles per second. Now that's lickity-split fast. Imagine if we could travel that fast. We could go from Houston to Rome faster than you can snap your fingers. Want off of the planet? Who doesn't these days? No problem. Ride the morning express to Mars in ten minutes".

"But here's the really fascinating part their measurements showed. Light speed didn't change, no matter how fast they were going relative to it. It's as if a friend drives past at 55 mph. Then you accelerate toward her to 40 mph, and she doesn't change her speed. But, son of a gun, you still think she is going 55 mph faster than you. This is not possible with cars at such slow speeds, but once you get to the speed of light, as light always will, it does happen. This was one of the cornerstones of Einstein's theory of relativity. Einstein clearly stood on Michelson and Morley's shoulders".

Science describes reality, at least as much as Mother Nature is willing to reveal, in terms we limited humans can understand and make use of. It doesn't attempt to describe one's wishes or political agendas or how one prefers reality to be. It describes what it is.

Through reality, science judges truth. Without it, anything can be claimed to be true, no matter how bizarre, with no way of refuting it. Flying monkeys, bending spoons with a thought, magnets curing cancer, aliens building the pyramids, ghosts haunting our homes. All are exposed as nonsense.

And it has served us well by uncovering some of the universe's secrets. This is one of the keys to its success since reality then challenges science, keeping it honest and giving it a roadmap. Drive off this path, and you are back to quackery and snake oil. And while you can fool some of people some of the time, you can't fool Mother Nature any of the time.

And that's it, really. That's all there is to it. Science has other attributes, but these are only to keep it honest and efficient. Science could survive without them, though perhaps not as opulently.

Here are a few:

Peer-Review

Researchers publish their work in journals, thereby publicly claiming credit and displaying their scientific prowess for all to see. Yes, it is somewhat like a peacock mating display, but do you think little peafowls would ever appear if their soon-to-be parents didn't go through this ritual? Same for researchers. No, not to produce little researches, but strutting their stuff will sire new research projects and opportunities for them. Well, maybe little researchers, too, but this is the wrong book for that.

Which then brings on the peer reviewers, the soup Nazis, who pass judgment on their work. Three or so experts in the field critique the work before it is set loose on the public. The peer reviewers ensure the work is of a high enough quality and accuracy to meet whatever standards the journal sets.

The peer review process has always been considered the gold standard of scientific review. It self-regulates the quality of any scientific work presented to the public. If it is not good science, it will either be rejected by the reviewers or subject to significant modifications. If allowed to be printed by the journal, the peer reviewers acknowledge it has met their standards, and as far as they can tell, it is worthy of recognition. More importantly, others should feel confident to use the results and build upon it in their own research. Plus, once it gets published, it is reviewed by everyone subscribing to the journal, oftentimes harshly by rivals.

An unfortunate example of violating the peer review process was the work on cold nuclear fusion by Fleishmann and Pons. They believed they had discovered a method of nuclear fusion energy production but at room temperature. If true, it would change the world and lead to cheap, unlimited, emission-free energy. Hard not to get excited about it.

But maybe they got too excited because they made the unforgivable mistake of going public with their results before it was properly peer-reviewed. And never wanting facts to block a good story, the press happily obliged. It turned out their work was flawed and could not be duplicated by experts in the field. They had not discovered cold fusion. It was an honest mistake but had they gone the traditional route of peer review; they would have been spared the embarrassment, which shows the power of peer-reviewed journals.

The peer review process applies only when researchers publicly disclose their work. Much science is hidden because it has value its owners wish to keep from competitors, be it other companies, countries, or foreign militaries. It is the marketplace (or the battlefield) that does the reviewing by deciding if it will make any money for the owners. This is a tough master, tougher than the peer reviewers, since the science has to work to be of value, but that's not enough. It also has to satisfy a commercial need. Further, it not only needs to be profitable, but it must be at least as profitable as other products the company offers or it will be quickly and mercilessly culled.

Some scientific advances never pass muster. The Concorde, a supersonic commercial jetliner, is an example of excellent scientific and engineering advances that the marketplace ultimately rejected as being too expensive for the flight time saved. It never made any money and was government subsidized for its prestige. Good science, bad politics.

The Edsel, a fine automobile from Ford, suffered from a seriously botched marketing campaign and lost $250 million in the late 1950s [1]. Good science, and really bad corporate leadership.

The invention of the wheel and harnessing fire may well be the two greatest discoveries ever, but there were no peer reviewers back then. It was true science, nonetheless as the evidence of their value was clear. The wheel required less work to move stones and fire made more foods digestible, greatly adding to the calories available to support a larger clan.

Reproduced by Others

Let's say you brag to your friends about a secret location where the fish are practically jumping into your boat. Is this another one of those exaggerated tall tales fishermen are known for? Easy to check. Your friends will go fishing there and will either come home with a supply of trout or hunt you down for chum.

It's the same with science. Can other researchers duplicate the work, or is it a tall fish tale? Once it is duplicated by others, it moves from a possibility to a confirmed discovery, and the scientific principles upon which it is based are deemed sound. If it can't be duplicated, all sorts of warning flags will pop up. It could be the discovery is a hoax, the other researchers trying to duplicate it don't have the necessary expertise or equipment, the inventor didn't disclose enough, and so on. No matter what, it needs further exploring as time, money, and reputations are seriously at stake.

Here's an example. Would you like to change someone's prejudices with just a twenty-two-minute conversation? A newly minted Ph.D. accomplished just that and produced data showing that having a short conversation with someone about same-sex marriage would convince that person to accept it. Further, this would also spread to their family and friends. As a result, he received an appointment at Princeton University as an assistant professor.

But alas, welcome to the world of fraudulent science. Did you really think Metabeliefs were so fragile that a simple conversation would wipe the slate clean? If so, then I have a bridge I'd like to sell you. Other researchers could not duplicate his results, which then led to further scrutiny, revealing his data was cooked [2]. That's a technical term meaning he was blowing smoke out his tailpipe.

At the least, this shows the power of independent experimental replication. And in case you were wondering, his Princeton appointment was retracted.

Experiments and Theory Go Hand in Hand

A full moon will cause strange behavior, even lunacy,[i] and there are anecdotal stories to support it. But how does the moon pull this off? Can you ascribe a theory to it? Can it be explained? It can't, which makes the anecdotal stories perhaps not disproven but highly suspect and more likely due to overcharged imaginations.

If data is science's master, then the theory is its teacher. All too often, people get hold of data and misuse it because they don't appreciate its context and limitations, since they lack a clear understanding of the science underlying it. A blind belief in data without understanding is a dangerous combination.

Microsoft learned this the hard way. They used artificial intelligence (AI) to train a social media chatbot [3]. AI computer software works by mimicking one of the ways the human brain learns by trial-and-error experiences. For instance, if AI is presented with many pictures of cats, it will learn what a cat looks like and will then be able to pick out a cat from a random scene most of the time. But the AI software has no understanding of what a cat is and how it determines that a picture is a cat and not a mouse. It is all experience and no wisdom. All data and no theory.

Microsoft used Twitter to train its AI. And everything the AI chatbot absorbed from Twitter it assumed was real and, therefore, could be safely used as a base for its chats with people on social media.

Twitter? Really? And they expected nothing would go wrong. Well, wrong it went, and in a big way. It lapsed into a Twitter storm posting over ninety-five thousand tweets in sixteen hours. That's almost two every second. But not just any tweets. Many of them were racist, anti-Semitic, misogynist, and who knows

---

[i] Guess where the word "lunacy" comes from? Luna is the Roman moon goddess.

what else. Microsoft mercifully ended the chatbot's brief life, but the damage was done.

The lesson? Be wary of anyone believing that piles of data will cure the world's ills. Only understanding will.

Testable Predictions

Scientists guess at the cause of something they see and then go about trying to disprove their guess. It's like believing the corner gas station has the cheapest gas. Then you drive around town checking all the other gas stations, trying to disprove your belief. Finding a single one with a lower gas price resets your theory.

How do scientists try to disprove their theories? An important element of all theories is that they must make predictions that can be tested to see if they are true or not. Newton predicted that two dropped balls of different weights would hit the ground at the same time. A bit counterintuitive as one would think a heavier ball would fall faster. It doesn't, and Galileo proved it by dropping cannonballs of different weights from the Leaning Tower of Pisa and observing that each hit the ground at the same moment. Admittedly, this is a bass-ackwards example, as Galileo preceded Newton by almost a century, but the point remains.

A famous example is Einstein's prediction that gravity will bend light. Now that sounds silly. Have you ever seen the beam from your flashlight bend toward the ground? I bet not, at least not when sober. But it was proven in 1919, which validated Einstein's' theory of relativity. To be fair to those sober-minded light seekers, it took the mass of the sun to bend the light and then by only a small amount.

But Science, Like Life, is Never that Simple

Science Approximates Reality

Many believe, with good reason, that science is fixed, and once a truth is discovered, it never changes. After all, truth is a description of reality, and what could be more enduring and eternal than reality? Water doesn't boil at 212°F one day, then 150°F the next.

So when science changes, as it often does, people wonder how that can be since the laws governing reality never change?

For instance, coffee was once thought to be harmful and could lead to ulcers and heart issues. As a result, we cut back on our morning cup of joe. Then along comes another study that said, nah, forget that previous study. Coffee may be beneficial after all.

Why can't these scientists get their act together? Should I or shouldn't I drink coffee?

The fly in the ointment is that we are too physically and intellectually limited to ever understand reality as it truly is, but rather, we see only some blurred approximation of it. It is not reality that is changing; it is our incomplete understanding of it, which we call science, that changes. The problem arises when we confuse science for reality and don't appreciate their differences.

A computer analogy may help drive home the point. A computer stores data, and it does so by flipping a switch to either on or off; 1 or 0, called a "bit" in computer parlance.[ii] Any item can be defined by a series of 1s and 0s, which computers can easily handle. The bit flipper is typically a transistor but could be anything that has an on-and-off state. Atoms and their constituents, even light, would work quite well. So if the universe is thought to be a large computer, very large at that, how much information could it store? Information that we believe to be the reality which we are eager to know and harness.

Jacob Bekenstein theorized that the amount of data any object could hold, be it a desktop computer or the universe, depends on how much mass it has and how big it is [4]. The heavier and the larger the more data it can hold. He then estimated the human brain could hold 26×10^{41} bits of information. That's 26 followed by 41 zeros. I didn't know I was that smart.

But how does that compare to the total amount of information out there? One researcher calculated that the universe holds 10^{120} bits of information [5]. That's 10 followed by 120 zeros. That's not to say all of that information will be useful to us; nonetheless, it is a measure of the entirety of reality. Now that's really smart, and it's 10^{78} times more information than we can ever know. An incomprehensibly large number, beyond anything we can visualize or grasp.

The universe dwarfs our minuscule bandwidth, making what we know essentially nothing compared to what *can* be known. In other words, most of the information the universe holds will forever be out of our reach. At least in our present form.

Accordingly, it's no surprise that we are constantly backtracking and updating what we think we know about science. We are stumbling through a dark forest of unlimited facts, occasionally finding something interesting and understandable but never seeing the whole. Then we bump into another piece of it and have a better idea of what it is, which upends what we thought originally.

Getting back to the coffee problem, later studies expanded on the original work and found many benefits to coffee the original study did not look for. It lowers cancer risk, gives some protection from Alzheimer's and Parkinson's disease, and so on. The newer studies were more robust and corrected for biases the original studies missed [6]. They stumbled onto a new piece of information that reality begrudgingly gave up, so they modified what came before.

Does that mean we can never know anything and all of science is as ephemeral as that fly doing a backstroke in the ointment? Of course not. Look around. We know quite a bit (or quite a byte? Sorry). And while it is insignificant in the bigger picture, we are equally insignificant, so it matches us quite well. Despite all our limitations, we have still managed to create civilizations, make tools, cure diseases, and feed billions of people. More importantly, despite the random picture I painted, we make progress each day.

Absent Absolute Truths, It's the Consensus of Experts That Rules

Science works by consensus since nothing can ever be proven to be absolutely correct or wrong. An everyday example of the necessity of consensus arises when a group of friends debates where to go for dinner. Each will defend their culinary preference, but it will come down to a consensus based on what everyone has shared because there can be no absolutely correct answer and the information about the restaurants is incomplete.

It's the same way with science since there are no absolute truths we can ever find. And because of our limited understanding, it always comes down to a consensus of the experts well versed in the topic to decide on what is true and what is not and which of the competing explanations is the correct one. There may be dissenters, but usually, the majority is correct, at least at that particular time and place. As new facts emerge from the ether, the consensus will drift to a new idea, if a bit slowly.

Global warming has long had such a consensus. Ninety-seven percent of climate experts agree that global warming is happening and that we are the culprits. This is as close as we can ever get to unqualified certainty on any human-explored topic. Within our human frailties, this means it is so. There can be no other reasonable interpretation.

Out of curiosity, what about the 3 percent dissenters? Benestad and his colleagues looked at thirty-eight recently published papers refuting global warming [7]. They found the work to be seriously flawed by cherry-picking the facts, violating laws of physics, and other shortcomings. Some mistakes were made by inexperienced researchers, and some of the papers were published in journals that would not

---

[ii] The letter c is represented as 01100011, the letter a as 0110000, and the letter t as 01110100. Therefore "cat" would be 01100011 01100001 01110100.

attract climate experts to peer review them. In other words, the consensus principle works.

But Consensus has a Downside

There is a downside to this. Getting back to food, my favorite topic—sometimes when the group chooses the restaurant, there is a single strong voice that sways the rest. Maybe they are charismatic or in some way a leader of our little party of eaters. So this superstar decides on what restaurant to go to, not necessarily by fiat, but by strong will and reputation. The others follow along.

And again, it is the same with science. Some prestigious and highly accomplished science leaders will drive the research direction and expect everyone to toe the line and follow their research agenda and techniques. Unfortunately, this can easily stifle new ideas.

Pierre Azoulay took the "when the cat's away, the mice will play" approach to examine this effect. He studied the aftermath when well-known science superstars died [8]. He examined 452 deceased superstars and discovered that their loyal followers greatly reduced their scientific output [9]. More importantly, newcomers increased their output in the superstar's field with impactful and highly cited work. With the superstar out of the way, there was no longer an alpha male protecting his harem, and a period of innovation and new ideas followed [10].

Peer reviewing in journals and grant money is another way of keeping the superstars unchallenged in their nest of dominance. Reviewers will always be influenced by the reputation and status of the applicant and will favor those having made significant advances. Success breeds success so the reviewers go with the low risk, safe choice with a proven team. They are likely to know and even admire the superstar which adds to the dominance effect. But they may completely miss the decline in the quality of their work as the superstars age and lose their mojo.

In a similar light, the rich get richer is also at play. Roger Penrose argues that once a theory becomes accepted and fashionable, young researchers are guided in that direction even when equally promising paths exist [11]. In a way, it is the path of least resistance since it is established that this accepted theory will not lead a young scientist's career down a dead end. It is the safe choice and it provides a well-developed map to follow. Plus, the superstars, ever ready to guide young researchers, will usher them into their way of doing things. That's a good thing since that's how neophytes learn, but when carried too far it produces imperfect clones when what is needed to advance science is Darwin's survival of the fittest.

By the way, superstars have produced significant breakthroughs during their reign that may not have occurred without them. That's why they are called superstars. But they may have dominated their field past their prime and, consciously or not, limited entry of other possibly younger rising superstars. They simply overstayed their cosmic welcome, but at least, in so doing, they inadvertently confirmed the wisdom of that aphorism "a need for new blood."

Follow the Money—Again

It is said that necessity is the mother of invention; what is most needed is what is researched. The wheel and fire discoveries are maybe the earliest examples of this. The heroic COVID-19 vaccine work is another. But if necessity is the mother of invention, then money is surely the father. Whoever controls the purse strings also controls which science problems are chased and what the superstars, and the not so superstars, will work on. One of the largest puppet masters pulling on the strings is the federal government. In 2020, it doled out $134.1 billion in research grants.[12]

You would think the most important science would be undertaken, and you would be right if important science is defined by Congress and its erstwhile lobbyists. They decide based on stakeholders, scientists among them, but also whomever they owe favors, which doesn't always align with the voters' wishes.

One way Congress circumvents the voters' rights is known as earmarks, which are specific projects written into Congress's budget that bypass competition and any semblance of fair play. They are usually given to Congress members' friends, important constituents, and big donors. It's where their friends slurp at the taxpayers' money trough while thumbing their porky noses at those who can't. How much? In 2020 it was $16 billion, spread among 274 projects [12]. That's $58 million per project for their friends and benefactors. Did you get any of it? I bet not unless you are among the 274 chosen.

The Bridge to Nowhere in Alaska is a classic one. It was an earmark costing taxpayers $300 million to connect two Alaskan cities, one with a population of eight thousand and the other with a whopping fifty people [13], even though an adequate ferry service was available. An outcry and some very bad publicity killed the bridge. But that didn't stop politicians from being politicians. The money was never returned to the taxpayers, as one might have logically supposed after the project was deep-sixed. It went to other Alaskan projects, so it remained an earmark after all, just dressed up for a different party.

Still, earmarks, also called plus-ups, do serve a vital function in our budgeting process. They provide consensus building, as any member of Congress with an earmark for their constituents will be much more inclined to work for and approve the federal government's annual budget [14]. In today's fractious and childlike partisan Congress, anything that brings these oil and water sides together is not to be scoffed at.

And the Bridge to Nowhere aside, most earmark projects are quite useful. Two examples: "Soil Carbon Sequestration Research Project" to Colorado State University and a CT scanner to St. Barnabas Hospital.

So there is some hope. Acknowledging global warming's influence, the Department of Energy reduced its spending on fossil energy research from $791 million in 2015 down to $562 million in 2020, a 29 percent reduction. And it was putting up $50 million for nuclear fusion energy research in 2020. Further, the Biden administration is pushing for $555 billion investment in climate, justice, and clean energy jobs [15].

Even with political headwinds resisting global warming, there is still progress being made. Hurray for us.

And Journalists Don't Always Get It

Most of us get our science news from journalists reporting on the web or in print, on TV, and so on. How accurate is the information we read or hear from these newshounds? It depends on a host of items, but one thing is certain: every time a piece of information goes from one person to another, it may get distorted. Remember the kids' game in which you whisper a comment to one person, who then whispers it to the next until the last person gets it. Often it doesn't sound anything like the original message. That's because each time a person passes it on, errors may creep in and these errors accumulate until the last garbled one is exhaled.

But there is more to it. Journalists don't always understand the subtleties of what is being reported. For instance, sometimes exploratory research is conducted in which connections are being sought between a disease and some lifestyle traits. It is meant only to get a feel of the land so as to inform future, more rigorous studies [16]. But the journalists will read the exploratory work, presume it is the gospel, and report it as a fait accompli. The above-mentioned coffee problem may have fallen into this trap. The first study was likely exploratory, not meant to draw hard conclusions but rather to give a general understanding and roadmap for others to follow.

Journalists will report on an article that appears interesting and newsworthy. But that's not how science works. Science does not live by one article alone. It grows by the cumulative knowledge that keeps adding to a particular topic over many years and by many researchers. A single result, while not wrong, though it certainly could be, is only a single result and is necessarily limited in scope. But journalists are myopic and only see the snapshot of that one article.

Journalists want a story that sells which means one that has a measure of sensationalism or controversy. It goes back to the newspaper adage a dog biting a man is not news, but a man biting a dog, now that's worth printing. People would not read the first but would line up in front of the newsstands for the second.

For instance, several news outlets reported that there was no evidence that flossing is helpful. The news outlets simply misunderstood the science reports [17], but needing a story and realizing a biting dog does them no good, they moved the narrative from flossing is good to flossing has not been shown to help. The media forced the research to fit their story line so it could sell. Irresponsible journalism? Probably.

And don't forget the journalists from the Ugly Table from chapter 3. Science from them? Bah. You're better off getting your science from the Saturday morning cartoons. At least it's entertaining.

On balance, most science reporting is quite good, especially if it comes from the Gold Standard Table. But being forewarned is being forearmed, so don't forget to floss and enjoy your coffee.

The Limits of Science

Despite its remarkable accomplishments, science has a limited domain, since it insists on working with only that which can be measured. There is much in the world that can't be measured: religion, philosophy, morality, and the arts, all of which are essential to being human, and science simply can't hang with them. While science is based on facts, religion is based on faith, the arts on emotions and beauty, and philosophy on truth and morality. We should celebrate what science gives us, but let's also be aware of what it can never be.

Connecting the Dots

1. Science describes reality and uses data and evidence for its validation. It has no time for "what we would like reality to be." It is only concerned with what is. It is the ultimate truth seeker.

2. Peer review is the gold standard of scientific review and self-polices itself to maintain its quality.

3. Another safeguard is the ability of others to duplicate the published work. If others can duplicate it, then it is likely valid and can be built upon. If not, then it's back to the drawing board, or worse, out to the woodshed.

4. While data and measurements are critical to scientific advances, equally important is an understanding of the data. The theory underlying the data must be known or at least approachable. One without the other is like yin without yang.

5. Reality is nearly unlimited, but our understanding is not so blessed. Science is only a rough approximation of what truly is. That is why science is constantly backtracking and tearing down and then rebuilding what it knows as it learns new things. It is not a fault of science but rather an acknowledgment of human limitations in what we can ever know.

6. That is why science can never be certain. It only deals with probabilities.

7. So when 97 percent of climate scientists say global warming is real, you can take it to the bank. Sure, there is a 3 percent uncertainty in this statement, but that is about as close as we can ever get to being certain about anything.

8. Most importantly, since knowing absolute truths is beyond our abilities, science relies heavily on the consensus of experts to judge the validity of any scientific theory.

9. Mostly this works. Sometimes it doesn't. Science superstars can warp science's direction the same way gravity can bend light. When a strong individual decides on a direction, consensus fails.

10. As is true in life everywhere, money talks. Whoever funds the scientific research also controls the scientific advances.

11. Be wary of journalists with tight deadlines reporting on scientific breakthroughs. Journalists operate in a sphere different from scientists and sometimes don't truly understand an advance's trajectory. Anyway, stick to the Gold Standard Table of news and we will be as safe as can be expected.

12. And always understand what science can never be. Its greatest asset is its greatest weakness. It relies solely on data and facts and so can never join the dance with religion, the arts, morality, all of which deals with intangibles so important to being human.

Afterthought: Resistance is Futile

Researchers at the University of Delaware figured out how to turn us into cyborgs. Scary but true. They discovered a polymer that adheres to the brain and allows electrical impulses to pass [18], which is exactly how the brain does its thinking. And they want to upload instructions into these imbedded polymers that the brain must follow.

But why stop there? In the interest of science they certainly did not. Because they've added the coup de grace by allowing dopamine to flow on demand. Dopamine is the brain's heroine-like drug that rewards us when our brain believes we've been good little boys and girls.

Displaying indifference worthy of the Guinness Book of Records, the researchers don't seem put off by what could possibly go wrong, clearly belonging to the "we'll burn that bridge when we come to it" school of thought.

But think what we've got here. This is great. We will finally rid ourselves of those annoying habits by those thoroughly annoying people. We can upload instructions into people's brains to eliminate talking in movies and putting ketchup on hotdogs. Can you say halleluiah?

I know this would never be used to harm anyone; when have we ever turned science against people? But let's let our imagination run wild for a moment. The U. of Delaware can turn all of us into automated drones to do their bidding. Now, that'll create an endowment to beat all. Step aside MIT and Harvard. After all, if these researchers are blind to what can go wrong, I'm sure they will giggle their way to the next logical, inevitable step. We will all work collectively like bees in a hive and believe whatever is programmed into us, no matter how absurd, and we will feel good about it to boot. Cell towers cause virus infections. Check. The moon landing was faked. Check. JFK was assassinated by the CIA. Check. High-flying jets are spraying chemicals to reduce the population. Check.

Uh oh, this sounds way too familiar. Isn't this how social media works? You bet. Sorry, Delaware, you're a neuron short and a synaptic late. Maybe you can invent something else, say turning people into Reptilians to control the world. But leave mind control to social media. You are getting sleepy . . . social media knows all. On the count of three wake up.

References

[1]   Edsel. https://en.wikipedia.org/wiki/Edsel/

[2]   Geggel L. Retracted study: Short talk can change people's views on marriage and same-sex couples. Livescience

[3]   Olavsrud T. 5 Famous analytics and AI disasters. CIO (Framingham Mass) 2020; 22. https://www.cio.com/article/3586802/5-famous-analytics-and-ai-disasters.html/

[4]   Bekenstein bound. https://en.wikipedia.org/wiki/Bekenstein_bound/

[5]   Minkle JR. If the universe were a computer. Physics 2002; 24. https://physics.aps.org/story/v9/st27/

[6]   a) What is it about coffee? Harvard Health Letter 2012. b) Does coffee offer health benefits? Mayo Clinic

[7]   Geiling N. Let's see what happens when this group of scientists retests studies that contradict climate science. Think Progress 2015.

[8]   Azoulay P, Graff-Zivin J, Uzzi B, *et al.* Toward a more scientific science. Science 2018; 361(6408): 1194-7.
[http://dx.doi.org/10.1126/science.aav2484] [PMID: 30237341]

[9]   Azoulay P, *et al.* Does science advance one funeral at a time? NBER Working Paper Series 2018.

[10]   Resnick B. Study: Elite scientists can hold back science. Vox 2015.

[11]   The folly of fashionable thinking. Science 2016; 354(6308): 7.

[12]   2020 congressional pig book. Citizens Against Government Waste

[13]   Dinan S. Alaska kills infamous 'Bridge to Nowhere' that helped put an end to earmarks. Washington Times 2015.

[14]   Broadwater L, *et al.* As earmarks return to congress, lawmakers rush to steer money home. https://www.nytimes.com/2022/04/01/us/politics/congress-earmarks.html/2022.

[15]   Cerceo E. The climate crisis disproportionately burdens women. Sunday Democrat 2022. Available From: https://www.nj.com/opinion/2022/06/the-climate-crisis-disproportionately-impacts-w-men-l-opinion.html

[16]   Gutting G. What do scientific studies show? New York Times 2013.

[17]   Gilmore E. Why trust science? Science 2019; 365(6457): 6.

[18]   Smithers D. Scientists discover new material that could see ai merge with the human brain. 2020.

CHAPTER 9

Nothing in Life Can Ever Be Certain. Nature Forbids It

"When you have eliminated the impossible, whatever remains, however improbable, must be the truth."

—*Mr. Spock quoting Sherlock Holmes*

It's true, we can only know the tiniest amount of reality's secrets, as the previous chapter has shown. But Mother Nature, not content with stopping there, has decided that the little we can see will be through fogged-up lenses. Adding insult to injury, she puts a hard stop on how accurately we can know anything, which no amount of science advances can ever get around. I mean, if you're pummeling someone, you might as go the whole nine yards. She's a tough old biddy, that one.

The limit is not just because of our small cranial capacity, though that doesn't help us any. It is also fundamental that all things can only be known with limited accuracy. For instance, flip a coin. We can never know if it will come up head or tails with certainty. We can only know that 50 percent of the time heads will show. And even that's not certain.

The problem is that reality doesn't itself know what it's doing. Ask any electron orbiting any nucleus to tell you where it is at any given moment. It can't. Not because electrons can't speak but because it has no definite location, just a likelihood of being here or there or somewhere else. In other words, it isn't anywhere at all, at least not in the sense of what location means to us. It is similar to the flipped coin, which can be in either of two states and we can never know which one it will be until we see it land.

Now multiply the electron's confusion by all the other electrons orbiting every nucleus of every atom and it's a wonder we even know our own names.

Uncertainty comes in two flavors. First is the plain vanilla kind due to our limited but growing knowledge and skills. There was a time when we didn't know where diseases came from and ascribed it to divine punishment for our misdeeds. Then civil engineers came along and in the 1800s cleaned up the town's water and sewerage and people's health and mortality greatly improved. People still got sick, so we were still baffled until bacteria was discovered, and in 1928, penicillin to kill the bacteria. And now we are struggling to understand and defeat viruses. Continual progress but always uncertainty.

The second and more fundamental flavor is that reality is not deterministic and does not give the same answer to a question asked twice. This was first noticed by researchers when they probed the subatomic level and developed quantum mechanics to explain what they saw. And it has been confirmed by countless experiments. Reality is just a never-ending bundle of probabilities. We are certain of things only because we are ignorant of what we can't see.

So how can we be certain about uncertainty when nothing is allowed to be certain? Yeah, I'm as confused as you are. But we will make sense of it.

CHAPTER CONTENTS

Vanilla-flavored Uncertainty is all Around us

Doctor: "I am sorry to tell you, but your mammogram tested positive."
Woman: "This is awful. Are you sure?"
Doctor: "These tests are 87 percent accurate."
Woman: "Those are terrible odds. It means I very likely have breast cancer!"

This fictional dialog probably takes place countless times, causing needless alarm to the patients and their families. Should she be worried? It is an emotionally and physically debilitating disease, so of course, there are reasons to be worried. But what are the true odds of actually having breast cancer with a positive test result?

Our fictional woman assumes, as most do, that it means she has an 87 percent chance of having cancer. That is, if one hundred women test positive from a mammogram, eighty-seven will have breast cancer. This is dead wrong and doesn't mean anything of the sort and is an example of important everyday events that have uncertainty we miss.

It is simple to unravel what's really going on and why the doctor and patient don't understand the uncertainty in what they believe to be true. Following Kit Yates's approach [1], let's assume 100,000 women take a mammogram test and 129 of them actually have breast cancer but don't yet realize it.[i] The first issue is What does the 87 percent accuracy really mean? The mammogram is looking for cancer cells. It either finds some or it doesn't. So the accuracy number means it will correctly find cancer cells in 87 percent of the women who have cancer cells to begin with. Thus, of those 129 poor souls, 112 will be correctly diagnosed by the mammogram.

Now comes the elephant in the room that no one acknowledges. Mammograms are imperfect, as we have seen, since they correctly identify only 87 percent of actual cancer patients. But they also err on the other side of the ledger and mistakenly call out healthy women with a false positive, incorrectly declaring they have cancer. And they do that 9.5 percent of the time [2].

That means mammograms wrongly proclaim "you have cancer" to 9,448 women who are, in fact, clean as a whistle, as compared to correctly identifying only 112 women.[ii] Do you see the elephant yet? So if you test positive, you might be part of the unfortunate 112 women, but far more likely, you will be part of the needlessly frightened 9,448 women.

In other words, if you test positive, your chances of having breast cancer are:

$$\frac{112}{9{,}448 + 112}\,100 = 1.2\%$$ (1)

That's 1.2 percent not 87 percent. Big difference.

Mammograms save quite a few lives, make no mistake about it. But they also give considerable angst and needless follow-up procedures from their false alarms; and make no mistake about that either.

A mammogram, like all other diagnostic tests, does not provide an absolute answer. Rather, it gives us valuable information, uncertain as it might be, in which we can make decisions based on our needs.

So let's understand what the mammogram test results really mean and not what others would have us believe.

The table below has a few more cancer screening tests using the same logic as for the mammogram tests above. The incidence is what our chances are, in general, of having that particular cancer. In other words, the risk in the general population for that particular cancer. The "True Risk" is our actual risk of a positive screening result when the false positives are also considered. For mammograms, the True Risk is almost ten times that for a woman in the general population but nonetheless still small.

Cancer Screening Outcomes from a Positive Result.

Cancer	Test	Accuracy (percent)	False Alarms (percent)	Incidence (percent)	True Risk (percent)
Breast	Mammogram	87	9.5	0.13	1.2
Ovarian	Ultrasound	99.8	0.01	0.009	61
Prostate	PSA	85	75	0.11	0.13
Lung	Low-Dose CT	88.9	7.4	0.047	0.56
Pancreatic	CT	81.4	57	0.013	0.020
Cervical	PAP	98	46	0.035	0.074

Notes: 1. The above table uses a number of references [3-8].
2. Accuracy is also called "sensitivity."
3. False alarms are the opposite of specificity and are also called "false positives." 4. Incidence is also called "prevalence."

So the question becomes, when the True Risk gives us any additional information above what we already know. In other words, how much higher is the True Risk when a screening test yields a positive result as compared to our general chances

---

[i] From the National Cancer Institute, the incidence of breast cancer in the general female population is 0.129 percent. So of 100,000 women, 129 will have breast cancer.
[ii] That number is determined by multiplying 9.5 percent times (100,000 - 129) which equals 9,448.

(incidence)? For ovarian cancer, a positive test result is significant. We go from an incidence of 0.009 percent to a True Result of 61 percent. But for most others, while our chances go up, they are still relatively small.

The PSA test for prostate cancer is the worst. It shows a True Risk almost identical to the incidence. It is useless. We know our risk of having prostate cancer is 0.11 percent because that is the risk in the general population. Telling us that a positive test result increases that to 0.13 percent tells us nothing at all. It provides no useful information and should never be used to make medical decisions.

Uncertainty rules, as it always will. But let's not be fooled by it either.

Beating the House

Everything you need to know about uncertainty can be learned by flipping a coin. Plus, it leads to two ways of beating the casinos.

To start, we can agree a flipped coin will have an equal likelihood of coming up heads or tails. We have no assurance of which one, except that one or the other must show its face. But what about in the long run? Will flipping a coin always result in an equal number of heads as there are tails? After all, each is equally likely. If there are thousands of heads, will there be an equal thousand of tails?

So I offer you this wager. Let's flip a coin two thousand times and keep track of the number of heads and tails. I'll give you a dollar each time their totals are equal and you will give me a dime each time they are not. I'm giving you 10 to 1 odds. Deal?

I'm sure you said no since by now you must realize I am not to be trusted. The number of heads and tails will almost never be equal. A little reflection will prove this point. Imagine we get to a point at which the sum of the heads exactly equals the sum of the tails. But what happens on our very next flip? It is no longer 50/50. Every time we hit the magical 50/50, our next flip deletes it. In fact, there are more flips that are not 50/50 than there are, and by a lot.

I simulated flipping a coin 2,000 times. The 50/50 mark occurred 138 times, just 7 percent. Nothing even close to the expected 50/50. So you raked in 138 dollars while I grabbed 1,862 dimes. And since my dimes add up to $186.10, I have taken from you a net $48.10. Thanks.

Now let's make some money from the casinos. We'll bet on black (or red) on a U.S. roulette wheel. There are an equal number of black and red numbers but that doesn't mean our chances of winning are 50/50. Just as in the coin flip, there is uncertainty waiting to snare us. There are also two numbers with no colors and we lose if the ball falls on one of them, just as we lose if the ball falls on the color not chosen. Thus, the odds of a roulette wheel win, while similar to flipping a coin, is worse at 47 percent.

Let's bet $10 on each turn with a payout winning of $10. The following figure shows our accumulated wins or losses with each simulated spin. We start out winning up to the forty-first bet, then we start to lose, recover, and then lose again.

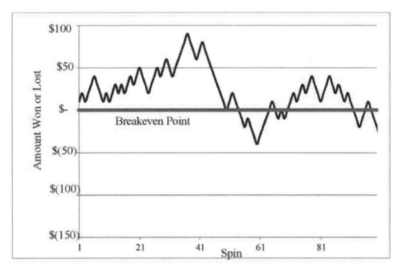

The next plot (figure provides a broader view with ten players). Some start out winning, but after forty or so spins, nearly everyone is in the hole. The small statistical advantage greatly skews the results in the house's favor.

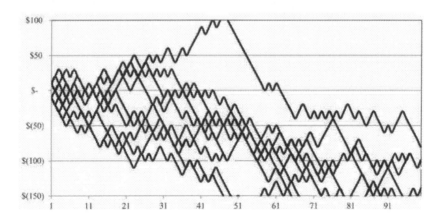

So how do we beat the house? Simple. If we are lucky enough to get ahead early, we take our money off the table and get out of the casino as fast as we can and don't look back. In the long run, we will almost always lose. Not with any certainty, as a few will make out okay, but most of us will be going home a good deal poorer. In this simulation, most will not get past forty spins, and many not even that far.[iii] No winners here.

And just as we were fooled into thinking a coin flip will average out to 50/50, so we are fooled into thinking a 50/50 black or red likelihood will result in anything but our money moving from our pockets into theirs.

Here's the second way to beat the house. Pick an amount you would be happy to win. Say $10. No point in being greedy. Place a $10 bet on black as before. If it comes out black, we take our $10 winnings and leave. Mission accomplished. But what if it comes up red? Then we double our bet to $20. If we win, we get back the first $10 we lost plus another $10 winnings, and again mission is accomplished, so we go home. But what if we lose twice in a row? Double our bet again. Keep doubling until we win our $10 and then leave. Simple with no downside. Well, almost.

The following figure shows another simulation of our accumulated fortunes if we continue to bet and double our bet after each loss.

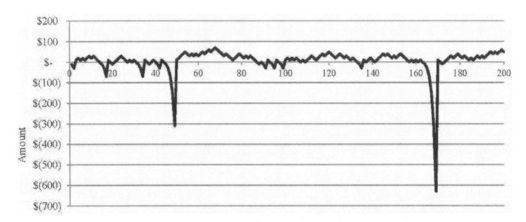

Look at what is happening. We hover around zero and make a small gain here and there. But look at the loss spikes. They are huge. One is sixty times the winnings and the other is thirty times. We are winning a small amount compared to losing a large amount. And if one of those spikes exceeds the casino's betting limit, we just lost our shirt. This only works if betting is unlimited, we have unlimited money, and we are willing to make a small amount of money with large bets.

[iii] This, and the following example, are simple computer simulations and may not reflect the real world.

Plus, the casinos are much smarter than most of us, so they will set the maximum bet maybe only ten times the minimum bet, so doubling a bet carries the real danger of running out of room before a win is assured, leading to one of those spike losses. Don't try it. In this case, the only way to beat the house is to not walk through the door. Stay home.

Our problem is that we have no intuition for probabilities. Our survival never depended on it. We are really good at pattern recognition, since quickly seeing the saber-toothed tiger crouching in the tall grass separated the genes of those who could from those who could not, which propagated to the next generation. But we were never at risk from an attacking baccarat table.

The gambler's fallacy is a well-known example of this. Sometimes black will come up several times in a row. The gambler will mistakenly believe that red is thus due for a turn, even though there is an equal likelihood of each one coming up on any spin. Flip a coin and it may come up tails four times in a row. What's the probability of it coming up heads next? 50/50, just as it is for tails.[iv]

The *Sports Illustrated* cover jinx is an offshoot of the gambler's fallacy. The magazine showcases on its cover athletes or teams having a recent exceptional performance. The urban legend is that this jinxes the athlete or team because their subsequent performance often falls short of their cover performance. No jinx here, just something known as regression to the mean. The exceptional performance was the athlete or team performing above their normal skills, which is why they made it onto the cover. Their subsequent downgraded performance is simply that they fell back to their average skills.

Kahneman pointed out a similar occurrence with flight instructors [9]. Israeli military flight instructors believed it was best to yell at student pilots rather than compliment them. They argued that when they complimented a student after a good performance, they would next do poorly. But when they yelled at a student who did poorly, they would next do well. In each case, the student merely regressed to the mean or went to their average performance after an unusually good or poor performance, as happened on the *Sports Illustrated* cover.

Simply put, we are not built to understand uncertainty and probabilities.

Still don't believe me? Okay, here's an experiment you can try then. Ask someone this simple question: Would you be more afraid of something that kills less than two hundred Americans or something that kills over five hundred thousand?

Easy choice, right? We would, of course, be more afraid of something that killed more people and we would choose five hundred thousand.

Next, ask them one final question: Which is to be feared the most—a terrorist attack or an automobile accident? I think most of us would choose the terrorist attack as the one to be most feared.

And now for the reckoning. Since 9/11, terrorist attacks have killed 173 Americans but auto accidents have killed 560,000. Yet most of us would have chosen the far less dangerous terrorists to fear. So our intuition is to irrationally fear that which is far less deadly. And not by just a little either.

Uncertainty rules us. We just don't realize it.

Fundamentally, Certainty Doesn't Exist and Never Will

Electrons are negatively charged and the nuclei they orbit are positively charged.[v] Since opposite charges attract, why don't the electrons spiral into the nuclei ending the universe and everything in it, including us? In fact, it would take only about twenty minutes for the entire earth to collapse into itself, forming a black hole [10].

The reason for our thankfully continued existence comes down to Mother Nature's uncompromising need for uncertainty. If an electron crashed into the nucleus, its location and speed would be exactly specified. The universe would then be far too certain for her liking. But first, a detour.

The master of uncertainty was Werner Heisenberg and his uncertainty principle, which states we can never know all we want about anything. No matter how smart we are, or how good our instruments are, or how hard we work. It is fundamental in the sense that science makes it clear some things will forever be out of our reach.

Specifically, Heisenberg realized the only way to explain the spooky world of quantum mechanics and particles on the atomic level was to acknowledge a limit on what can be known about both an electron's location and velocity. The more accurate we know one, the less accurately we can know the other. If we double the accuracy of one, we halve the accuracy of the other. Essentially, the electron's velocity and location are connected in a fundamental way such that the accuracy of its speed causes a slight wobble in its position.

The cause of the electron's fuzzy behavior is that it is really two beings in one, somewhat analogous to our System 1 and 2. It is both a particle and a wave. Double-slit experiments showed this bizarre duality of subatomic particles and their need to obscure their whereabouts.

[iv] Although Howard Beroff suggests the best strategy might be to ride black, as there may be a mechanical problem with the wheel skewing the results or the coin may be unfairly weighted for tails (Beroff, H., "How Logic Is Misrepresented in the Media and in Everyday Life," PowerPoint slides in Raritan Valley Community College Course on Statistics).

[v] That's what an atom is: a nucleus of protons and neutrons orbited by electrons.

Fire electrons at a wall. They will show a classical scatter pattern around a central point that you are aiming at. So far so good.

Now for fun. Place two parallel slits closely spaced to one another in the electron's path. What happens? If you were firing tennis balls instead of the electrons, some would get through each slit and you would have two bull's-eye patterns on the wall. Not so with the electrons. Each electron travels through each slit simultaneously. And each interferes with itself, whereby its peaks and valleys combine to produce a classical wave pattern. In the first instance, without the slits, the electron acts like a particle, and in the second instance, with the slits, it acts as a wave. Such is the peculiar life in the fast lane of the subatomic world.

This poor schizoid electron. Sometimes it behaves as a wave and sometimes as a particle. At the beginning of this chapter, it was clear the electron never knows where it is, and now it doesn't even know what it is. Talk about having a bad day. And just think, all mass is made up of atoms, all of which have many electrons.[vi] Everything with mass throughout the universe has this same uncertainty.

So because of its wavelike behavior, an electron can be in many places simultaneously, just as an ocean's wave hitting a beach is.

Now getting back to the reason our universe has not annihilated itself. If the electrons approached the nucleus, then their location would be specified exactly with no uncertainty and therefore, according to Heisenberg, their velocity could be anything, even infinite, since there is no bound on its error. But nothing can go faster than the speed of light (the universe's absolute speed limit) so the electrons can never get that close so as to keep from exceeding this unexpendable speed limit. It would be breaking too many of Mother Nature's laws, so it doesn't happen.

This is also why a true vacuum can never exist. The Heisenberg uncertainty principle also applies to energy and time [11]. If one is highly constrained, the other cannot be. So for very short times, the energy can be uncertain, leading to pairs of particles briefly materializing before they annihilate each other. Curiouser and curiouser.

Thank goodness for whoever invented the Heisenberg uncertainty principle. Once again, He (She?) saved our bacon.

This probability of where an electron can be, leads to more weird weirdnesses. Since an electron can be anywhere (though with less and less probability as it moves farther away from the nucleus) it could pass through a solid barrier. Not a likely occurrence but because there are so many electrons, some will.

[vi] The lawrencium atom, king of the electron holders, has 103 electrons orbiting its nucleus. Now that is one really confused atom

This is the basis of tunneling microscopes. By applying a voltage between the instrument and the target surface, some electrons will tunnel to reach the surface and by measuring the current fluctuations caused by the electrons, the target surface can be mapped to a much higher resolution than other types of microscopes.

"Here's an interesting Aside. Have you ever visited another planet or come in contact with an alien being? Maybe you have without ever knowing. Electrons have only a probability to be at a certain point in an orbit around their nucleus. Which raises an intriguing possibility. All of the atoms that make up your physical being have electrons. According to Heisenberg, at any time one or more of them could easily slip outside your body. In fact, one of them can suddenly appear across the galaxy. Highly unlikely but not prohibited by science, so it is possible. Who knows, some part of you might be visiting an alien world million of light-years away, canoodling with other like-minded electrons".

"Is it likely? Who cares. It's possible and that's the fun of it".

But what if we could know the precise location and speed of every particle despite Herr Heisenberg? Would it do us any good? Maybe not. Here's a thought experiment proposed by Richard Feynman [12]. Suppose we could measure the exact location and speed of every molecule in the room with unlimited accuracy. Yes, I know it is quite impossible and Heisenberg would be rolling over in his grave, but don't forget this is a thought experiment in which anything goes for the sake of visualizing a difficult problem. Let's also assume we have the mother of all computers and we could calculate the next position and velocity of each of those molecules with absolute certainty. We could then know the second position and speed of each molecule, then the third, then the fourth, and so on, for as long as we cared to. In fact, we could determine the location and speed of each of those molecules for all eternity. Trillions upon trillions of them. Nice!

Now comes the interesting part. Take just one molecule of those trillions upon trillions of molecules and say we have an error of this one measurement of one part in a billion in its position. Nothing more. What do you suppose will happen? This rogue molecule bumps into an adjacent molecule, so this second molecule now has that same error. The two molecules hit two more adjacent molecules, then those four hit four more, then those eight hits eight, then sixteen, then thirty-two, and so on in a geometric progression. And just like a virus, the error spreads throughout the population of molecules. In no time at all, seconds or minutes, complete chaos will reign, and we will have lost any knowledge of where or how fast any of those molecules are going. All of our impossible Herculean efforts to measure each one, all of that impossible effort and accomplishment, would be completely wasted. The position and velocity of any one of those trillions upon trillions of particles would now be completely unknown, despite the superhuman ability to measure each one exactly but one.

Certainty is an illusion and science, like everything else in life, can only deal in likelihoods and never in certainties. From our own everyday experience, we can measure how tall our youngster is by having her lean against a ruler on a wall. But our measurement can never be certain as the ruler is not exactly made and we will only be able to read it maybe down to an eighth of an inch. Beyond that, we have no knowledge of our daughter's height.

The difficulty arises when we demand the world around us to be sensible and stable and we believe that any uncertainty must be due to our own inadequacies. Not true. Not at all true. The universe is nothing more than a bundle of indeterministic uncertainties. It's not us, it's the universe just being itself.

And So, The Science of Global Warming Must Also Be Uncertain. Nothing Else can be Expected

Our understanding of global warming is uncertain in the same fashion as the examples given in this chapter, as everything in life must be. We will never be certain that global warming is happening and that we are responsible for it. Why? In science, absolute truths don't exist and never will. The IPCC acknowledges this as they give their conclusions in terms of "confidence levels," never in terms of certainty. Even their highest rating, "Very high confidence," has a nine in ten chance of being true. These are qualitative approximations only, but it drives home the point that global warming is highly likely, though not certain.

But that is no reason to reject it. It is simply a fact of life. The real question then becomes, Which is the less uncertain: global warming or no global warming? Here we stand on firm ground. The evidence overwhelmingly shows global warming is happening and we are the reason, whereas its opposite of no global warming has little to offer in its defense.

In other words, uncertainty or not, the Balance of Facts Principle, from chapter 3, rules. And of that you can be certain.

Connecting the Dots

1. Certainty is an illusion and can never be achieved.

2. Uncertainty, its evil twin, is everywhere. There is no avoiding it.

3. First, there is the vanilla kind of uncertainty due to our limited but growing

knowledge. Simple everyday events demonstrate this: mammograms, flipping a coin, and placing bets at a casino are examples of our inability to know, or even acknowledge, the uncertainty built into our lives.

4. But more fundamentally, uncertainty must exist, and no matter how far science advances, it will never advance far enough to eliminate it.

5. Science deals in likelihoods and never in certainties. It can't, since certainty, as elusive as that pot of gold at the end of the rainbow, does not exist in our universe.

6. But uncertainty doesn't mean useless. Do we truly know the details of how a child is conceived and grows in the uterus before being born? There is uncertainty in our minds, but such a miracle happens every day.

7. The Balance of Facts Principle drives science, allowing it to achieve greatness, despite the ever-present uncertainty.

8. Global warming is irrefutable. Not because such a fact is certain but because it is overwhelming more likely than its opposite.

Afterthought: Sherlock Holmes was Wrong

Sherlock Holmes's quote at the beginning of the chapter can never be realized. Had Mr. Spock thought about it for more than a single episode, he would have admonished Captain Kirk:

"I am afraid, Captain, that statement is . . . quite . . . illogical. It presumes perfect knowledge of all possibilities, which is not only impractical but also theoretically impossible."

And since you don't argue with a Vulcan any more than you tug on a Klingon's cape, a better though less eloquent quote, which Mr. Spock, with his newfound human wisdom, would embrace is:

"When you have eliminated the impossible, and only the improbable remains, dig deeper. You clearly missed something."

Reference

[1]   Yates K. The Math of Life & Death. New York: Scribner 2019.

[2]   Accuracy of mammograms. https://ww5.komen.org/BreastCancer/AccuracyofMammograms.html/

[3]   Cancer stat figures. National Cancer Institute. https://seer.cancer.gov/statfacts/index.html/

[4]   New blood test for prostate cancer is highly accurate and avoids invasive biopsies. https://www.google.com/search?client=firefox-b-1-d&q=accuracy+of+psa+screening/2019.

[5]   Toyoda Y, Nakayama T, Kusunoki Y, Iso H, Suzuki T. Sensitivity and specificity of lung cancer screening using chest low-dose computed tomography. Br J Cancer 2008; 98(10): 1602-7. https://www.ncbi.nlm.nih.gov/pmc/articles/PMC2391122/ [http://dx.doi.org/10.1038/sj.bjc.6604351] [PMID: 18475292]

[6]   Costache MI, *et al.* Which is the best imaging method in pancreatic adenocarcinoma diagnosis and staging—CT, MRI or EUS? Current Health Science Journal. www.ncbi.nlm.nih.gov/pmc/articles/PMC6284179/#:~:text=Computed2017.

[7]   Chang L. Lung CT Scans Produce False Alarms. WebMD Cancer Center. https://www.webmd.com/lung-cancer/ news/20090601/lung-ct-scans-produce-false-alarms/2009.

[8]   Maxim L, *et al.* Screening tests: A review with examples https://www.ncbi.nlm.nih.gov/pmc/articles/PMC4389712/2014. [http://dx.doi.org/10.3109/08958378.2014.955932]

[9]   Kahneman D. Thinking, Fast and Slow. New York: Farrar, Straus and Giroux 2011.

[10]   Seigel E. What would we experience if earth spontaneously turned into a black hole? 2020.

[11]   What is the heisenberg's uncertainty principle? The Guardian.

[12]   The Feynman Lectures on Physics. Addison Wesley 1965; vol. 3: pp. 2-9-2-10.

CHAPTER 10

If the Military is Concerned, We Should Be Too

"It is the sense of Congress that climate change is a direct threat to the national security of the United States."

—*National Defense Authorization Act for Fiscal Year 2018*

What if Russia destroyed our missile interceptors in Alaska, China disabled our radar installation on the Marshall Islands, and homegrown terrorists attacked the Norfolk Naval Shipyard? What odds would you give for that amount of havoc? I bet quite low. But if global warming replaces the aggressor in each of these examples, then the odds are not just high, but a sure thing. A sucker's bet even.

The military knows this and their position is unmistakable as the Pentagon has stated:

"Climate change poses immediate risks to national security and will have broad and costly impacts on the way the US military carries out its missions"[1].

Consider that the military is tasked with protecting our country from outside enemies and have stated three pillars to accomplish this [2]:

• Protect the homeland

• Build security globally

• Project power and win decisively

Sounds simple, perhaps, although not easy to accomplish with the large array of conventional and nonstate enemies squaring off against us. But now U.S. armed forces are also facing global warming, an additional enemy with an arsenal vaster than all of the above combined. For instance, according to NASA, a hurricane will release the energy of ten thousand nuclear bombs during its life cycle [3]. Not big enough for you? Global warming is making them much more powerful still.

Global warming is a threat no different than Russia, China, or terrorists, homegrown or not. The result is certainly the same. Global warming will cause destruction, casualties, and economic decline, similar to an attack, with the only real difference being global warming will take a bit longer to achieve it.

And while the military's budget is substantial, it is not unlimited, which compels them to use their resources with little wasted motion. But global warming is forcing this substantial, but limited, budget to be spread thinner, taking assets away from their known and traditional adversaries. It's a zero-sum game. If you move some assets from here to there, you then have fewer assets here.

The military would never squander precious resources any more than a hoarder would give up their years of accumulated newspapers. So if the military is moving money from their typical threats, we can be sure they believe global warming is real and menacing. And if they believe it, so should we.

CHAPTER CONTENTS

Our National Security is at Risk

Global warming will lead directly to organized violence, and given we are the seven-hundred-pound gorilla on the world stage,[i] much of it will be directed toward us. And that which is not directed toward us will pull us in any way as it will likely affect our strategic interests.

Research shows violence increases with rising temperatures. For instance, from a review of sixty studies, rising temperatures due to global warming are projected to increase the number of violent conflicts by 50 percent by 2050 [4].

To put this in perspective, in 2019, there were forty-five conflicts around the world, two examples of which are the ongoing Syrian civil war and the Afghanistan [5]. So in 2050, there will be those wars or similar ones, plus another twenty-three piled on top for a total of sixty-eight global conflicts. And we may be involved in many of them. That's spreading resources far and wide.

Making matters worse, Syria and Afghanistan, lacking an effective central government, allow terrorists to gain a foothold and operate with relative impunity [6]. As general Anthony Zinni stated, "It's not hard to make the connection between climate change . . . and terrorism" [7]. Indeed, Hezbollah, al-Qaeda affiliated groups, and even some remnants of ISIS operate in Syria. In Afghanistan, the Taliban, al-Qaeda, and others operate. These are groups that would like nothing more than to harm the U.S.

Syria is a good example of global warming's cascading events that lead to direct threats against the U.S. Global warming made Syria's already existing droughts deeper and longer, which led to crop failures and livestock deaths, which were then exacerbated by the baffling inaction of Syria's central government [6]. This forced eight hundred thousand people to abandon their farms and move into already overcrowded cities, putting untenable pressure on their resources. Protests began, especially among people with nothing to lose. Then, baffling again, the government started shooting protesters, which triggered their civil war. Ever ready to step into a vacuum, ISIS found a place to germinate and terrorized the rest of the world until being mostly dismantled by U.S. and allied forces, though at a substantial cost, almost $11 billion as of 2017 [8].

One of many domino effects with us standing beneath the last one.

Failed states without a functioning government always lack money to fix their problems. That's why they are failed states. So even small problems will lead to their swan song. Global warming is no small problem, so a flock of swan songs is flying at them.

African countries are examples. Poverty is widespread, and their governments are often too weak to cope. Religious and ethnic stress and corruption move them from dire to an overwhelmingly impossible position. And an exploding poor population of 1.2 billion is expected to double by 2050 [6]. That's mind-boggling. Countries that God has seemingly forgotten. Their suffering just never ends, making them breeding grounds for terrorists the U.S. needs to prepare for.

And this is just the tip of the heat wave, since global warming is going to get worse, leading to not only more conflicts but also more terrorists willing to take advantage of the misery and suffering of those affected so they can apply their own brand of misery and suffering.

More specific threats are looming from Russia. The Arctic is warming twice as fast as anywhere else on the planet, making it more accessible to vessels. As Michael Klare points out [9], the melting Arctic will create a whole new ocean and another battlefield for the U.S. Navy. This will only add to the military's burdens to assure stability and access. And Russia has a sizeable Northern Fleet there, whereas the U.S. does not. And to make matters worse, according to a "Worldwide Threat Assessment," Russia seeks to influence all regions of the world and is adhering to a global grand strategy to be recognized as a great power [10].

Then there is China, that six-hundred-pound gorilla. China is using global warming to extend its influence [11]. Look at what they are already doing. They are claiming the South China Sea as their own and built a military artificial island on it. Not surprisingly, the South China Sea is an economic bonanza with large oil and natural gas reserves beneath the seabed, lucrative fishing, and $3 trillion of shipping passing through it each year [12].

China is also ramping up military provocations against Taiwan. And with an unabashed dictatorship, their leadership faces few checks and not much to rein in their power.

North Korea, no slouch in causing problems, is perennially on the brink of collapse and when it finally implodes, it will have global geopolitical reverberations. Their record-setting temperatures and droughts are putting over 10 million people at risk of malnutrition and possibly starvation [13]. As global warming picks up speed, things will get worse for them. A mass exodus north into China due to famine, putting even more pressure on China's global ambitions, is not unthinkable. "Not unthinkable"—now, there's a sad if not alarming phrase. Unfortunately, many situations once thought impossible have descended to "not unthinkable" soon to descend further to "unavoidable." Global warming has that effect on the words we use.

---

[i] We used to be the eight-hundred-pound gorilla, but China, maybe at six hundred pounds and getting fatter, is gaining ground.

Global warming doesn't always cause wars directly, but it heats up the powder keg many countries are sitting on, making violence more likely and perhaps pushing a marginally stable country over the edge. Case studies show global warming will cause other issues heaped upon existing ones, such as water scarcity and food shortages, which may lead to conflict even among countries not thought to be at a substantial risk [14].

Admiral Samuel Locklear stated that Middle East fights over land, food, and water are leading to unrest in addition to the usual religion or politics [15]. He further states "The real threats to global security today are climate change, population growth, water shortages, rising food prices, and the number of failing states in the world." Religion and politics light the fire. Global warming adds the fuel.

Military Bases are Under Threat

Getting back to the three threats mentioned in the beginning, maybe the Russians aren't coming, but melting permafrost is [16]. Add to that melting ice, rising sea levels and eroding shorelines, and the military's assets in Alaska are in great peril.

China invading the Marshall Islands? No need to. Rising sea levels and storm surges will do the trick without ever lifting a chopstick. It took $1 billion to build the installation, and at just six feet above sea level, it is expected to be underwater within the next twenty years. As John Conger, former acting assistant secretary of the Department of Defense (DOD) for energy, stated, "[DOD] is increasingly cognizant of the threat of sea level rise on its installations" [17].

As for the Norfolk Shipyard, Lt. General David Berger said, "The two biggest challenges are the rising water levels and severe storms that roll up the coast and through our bases and stations," which could seriously disrupt military activities there.

With seven thousand bases, installations, and other facilities [18], the U.S. is one of the largest landowners in the world. In the U.S. alone, two-thirds of the installations are prone to flooding, with thirteen having a history of flooding [19]. For example, it cost $5 billion to rebuild Tyndall Air Force Base after Hurricane Michael in 2018 [20]. Further, hurricanes and flooding have caused $3 billion damage to Camp Lejeune, North Carolina, and $400 million damage to Offutt Air Force Base, Nebraska [21].

Even our troops are at an increased risk. Seventeen troops died from 2008 to 2019 during hot weather training. In 2008 1,766 cases of heat stroke and heat exhaustion occurred, rising to 2,792 in 2018, a 60 percent increase over the decade [22]. Weather-related health impacts cost the military $1 billion from 2008 to 2018.

Humanitarian Missions will be Compromised

The U.S. is the largest single provider of humanitarian assistance worldwide, at over $8 billion in the fiscal year 2018 [23], with the U.S. military as the first responder in relief efforts.

With global warming increasing the number and severity of storms, the military's humanitarian burden is going to increase substantially. A good example was Superstorm Sandy in New York and New Jersey in 2012. The military provided over twenty-five thousand troops to help [24].

Catastrophic weather events, ever-increasing, will lead to even more demand on an already thinly stretched military.

The Insurance Industry is Worried about Mounting Losses

Whereas the military is mostly concerned with security, the insurance industry is mostly concerned with risk. They lose if their insured properties are damaged, and they win if these properties are protected. So, it is no surprise they are as concerned about global warming as the military.

"Follow the Money" is a major theme of this book, and at $630 billion in annual sales [25], the insurance industry is a rabbit to chase. Further, since 1989, weather-related natural disasters have caused losses of $4,200 billion worldwide and $2 billion in the U.S [26], and killed nearly a million people throughout the world. You can be sure the insurance industry is taking notice.

And they fully embrace the IPCC's conclusions on global warming [27], which will increase risks from hurricanes, storm surges, and flooding. Lloyd's of London views global warming as the industry's most urgent problem [28].

Insurers detest risky risk. That is, a risk that has its own uncertainty. For instance, they can determine the likelihood of a one-hundred-year flood and therefore set their premiums to assure a profit despite some guaranteed claims. But when the risk itself cannot be accurately predicted, they are left holding a bottomless cup of cash payouts. How do you set the premium of a hundred-year flood that occurs whenever it chooses?

The risk greatly depends on how people react to global warming, which is as unpredictable as the path of a drunkard's walk. Will we embrace global warming issues, or will we remain a nation divided, driven by social media, tweets, and cultural shock jocks? Who knows?

Plus, there is much left to learn about what controls the number and intensity of hurricanes, their major boogeyman. Global warming will increase all of these but by how much? How do they price it? How do they make money? How do they avoid bankruptcy when they are overwhelmed with claims? When do they fold their tents and let nature have its way with their ex-customers?

Add to this, the technologies being contemplated to combat global warming come with their own suite of risks. Technologies such as electric cars are safe enough, though how many will be sold is up in the air. But what about geoengineering technologies we might be forced to deploy? Sprinkling the air with magic fairy dust that blocks sunlight could very well turn into the eighth biblical plague when overwhelming droughts and floods hit different parts of the world. Yet another insurer's nightmare.

What frightens insurers the most are the tipping points; when an otherwise contained global warming problem suddenly and without warning triggers an avalanche-type crisis. Permafrost thawing and its possible eruption of massive amounts of carbon is one example of keeping them up at night. Increased wildfires are another, as evidenced by recent ones in Australia and California. Fires result in insurance losses, but they also add CO_2 to the air, which makes global warming worse, leading to a death spiral of more losses which then leads to more CO_2 and down they go.

The speed of global warming change is also critical to insurance companies, just as it is for animals and plants facing extinction. A slow increase in temperature and other related problems gives the insurance companies time to learn and adapt. But when it is rapid, they are no better off than the long-gone Dodo bird.

Even the Financial Folks are Hedging their Bets

Investors consider global warming-induced droughts and wildfires to be more of a problem than turbulent markets, cyber-attacks, or geopolitical instabilities [29]. And, their three top concerns are extreme weather, migration caused by global warming, and natural disasters.

The hedge-fund billionaire Thomas Steyer commissioned a study that concluded that business leaders are not adequately focused on global warming impacts on finances. Robert Rubin (former treasury secretary in the Clinton administration) said, "There are a lot of really significant, monumental issues facing the global economy, but this supersedes all else [30]".

Investors are demanding companies evaluate and plan for the risks of global warming. According to Alison Martin, chief risk officer at Zurich Insurance Group, investors don't care about the manager's personal beliefs about global warming. Their message is clear. Global warming is real and you had better prepare for it irrespective of your politics. Beliefs have little currency when large amounts of money are at stake. Only facts and projections matter.

Plus, investors, along with shareholders and regulators, are pressuring companies to disclose the adverse financial impacts from global warming, such as supply chain disruption, stricter climate regulations, and the declining value of fossil fuel investments [31]. How important is this? Consider that five hundred of the biggest companies face $1 trillion losses from global warming, which could materialize in the next five years. Issues such as supply chain disruption (Hitachi) and increased costs of cooling energy-burning computer centers (Google) are just two examples. The largest companies estimate that at least $250 billion in assets may need to be written off as they will be useless. This includes buildings that become flooded and power plants that cannot meet new pollution regulations.

Moody has warmed coastal cities that their credit will be downgraded if they don't prepare for global warming and take steps to deal with risks from surging seas and intense storms. Texas, Florida, Georgia, and Mississippi are the most at risk [32].

And if global warming isn't bad enough, as taxpayers, we are going to get sucker punched by banks. Banks are offloading mortgages to the federal government that is adversely affected by global warming, such as those likely to be flooded. And with $60 billion to $100 billion in coastal area mortgages issued each year [33], that's a lot of potential unloading. So when these mortgages default, you and I pay for them.

Remember the 2008 banking crisis? That's when banks and other lenders targeted low-income home buyers to buy mortgages they could not afford, using questionable predatory practices. The bubble collapsed, the loans defaulted, the banks used their considerable influence on lawmakers, and the U.S. government bailed them out. Guess where the government got the money for the bailout. Taxpayers: you and me. And now the banks are at it once more. Or as Yogi Berra said, "*It's déjà vu all over again.*" All because of global warming.

Connecting the Dots

1. The military's world is an unforgiving one. Conspiracy theories and lies don't survive on the battlefield.

2. And they realize that global warming is real and is hindering their ability to protect our country.

3. Our national security, their chief concern, is always under threat, but now global warming is multiplying that threat.

4. Global warming will siphon off a good part of the military's budget, leaving them with considerably reduced assets to protect us against our traditional enemies.

5. As global warming pushes marginally stable countries over the edge, terrorists will gain a foothold to launch attacks against the U.S.

6. Russia and China are taking advantage of the mayhem global warming is causing.

7. As the Arctic ice melts, Russia is moving its Northern Fleet to take advantage. Yet another battleground for the U.S. Navy, requiring more assets.

8. U.S. military bases are under siege from global warming as sea levels rise and storms become fiercer. More money will be needed to fix them, but more importantly, once a base goes underwater there is no recovering it. It is an asset gone forever.

9. Insurers are afraid of mounting losses due to global warming and fully accept the IPCC's conclusions.

10. Even the financiers are running scared. Many believe global warming is the most significant threat to the financial markets.

11. Don't believe global warming is real? Just ask those with the most to lose from it: the military, the insurers, and the financiers.

Afterthought: Long Shots aren't what they used to Be

There was a time when you could count on hundred-year floods occurring every one hundred years. That's why they were called hundred-year floods. Things were sensible back then. Not anymore. Now, hundred-year floods occur every twenty years [34].

Ten-year floods? I guess they will occur every two years. Yearly floods, every few weeks? Next you're going to tell me the tooth fairy is straight and pancakes are really flapjacks.

Global warming is sneering at weather statistics. The hottest year on record is usually the current year. Now that's an unsettling trend not at all pleasant to think about. Extinctions are erasing more species than at any time in humanity's past and maybe the most our planet has ever experienced.

Global warming is also upending the power brokers, those who run our country both on the surface and beneath. Let's shed a tear for the WASPs' fading power.[ii] No not the stinging kind. The blue bloods who have controlled our country since King George and the Brits gave it up and decided to focus instead on their monarchy and those nostalgic days of glories past. But the Brits did give us the Beatles, the Rolling Stones, the Who, Black Sabbath, Queen, Led Zeppelin, and dozens of others. The sun never sets on the British Empire's rock bands.

So who will be stepping into the power vacuum? Lean in a bit closer. I don't want this to get out. You can keep a secret right? I have no idea. No, that's not the secret. But I know who does know. Global warming will be the new kingmaker although maybe it's better thought of as the new Grim Reaper maker.

Sure, old white working-class males are trying to hold on to the illusion of power they never had, while the inexorable march of demographics is changing the U.S. political landscape. Then there's money, power, and most cursedly, those obsessively desperate for both at any cost. Global warming is making all of that as irrelevant as invisible ink tattoos.

So here's the secret. If we don't act forcibly and quickly, global warming will knock the military, the insurance industry, the financial markets, and much more on their ears.

And when they go down, so do we. A long shot? Not anymore.

[ii] WASP stands for white Angle Saxon Protestant for those who haven't figured out who is running our country and have been since the *Mayflower*.

Reference

[1]   Climate change threatens national security says pentagon. United Nations Climate Change 2014.

[2]   National security implications of climate-related risks and changing climate. 2015.

[3]   How much energy in a hurricane, volcano, and an earthquake? howstuffworks.

[4]   Landau E. Climate change may increase violence, study shows. CNN 2013.

[5]   List of armed conflicts in 2019. Wikipedia.

[6]   Holland A. Preventing tomorrow's climate wars. Sci Am 2016; 314(6): 60-5.
        [http://dx.doi.org/10.1038/scientificamerican0616-60] [PMID: 27196845]

[7]   Persuading conservatives how to reach conservatives on climate. Climate Chat.

[8]   McCarthy N. The cost of the air war ageist ISIS has reached $11 billion [Infographic]. Forbes 2017; 1.

[9]   Klare MT. All Hell Breaking Loose. New York: Metropolitan Books 2019.

[10]   Shapiro D, Yefimova-Trilling N. Worldwide threat assessment on russia: 2019 vs. 2018, Harvard
        Kennedy School Belfer Center, 1 February 2019; Doffman, Z., Pentagon Report Warns on Threat to
        U.S. from Russia's Dangerous Global Influenc. Forbes 2019; 1.

[11]   Conger J. U.S. Navy and marine corps nominees highlight the threat from climate change. Center for
        Climate and Security

[12]   South China Sea. Wikipedia

[13]   Burton J. Climate change and North Korea. Korean Times 2019.

[14]   National security implications of climate-related risks and a changing climate. 2015.

[15]   Four-star admiral admits what scares him most: Climate change. Yahoo News.

[16]   Nazarayn A. Climate change could lead to global catastrophe, new report warns. Yahoo News. 2020.

[17]   Mooney C, Dennis B. The military paid for a study on sea level rise. The results were scary.
        Washington Post. 2018.

[18]   Climate change adaptation roadmap. Department of Defense 2014.

[19]   Developing adaptation strategies to address climate change and uncertainty. SERDP, SERDP and
        ESTCP Webinar Series 2019.

[20]   Banerjee N. U.S. intelligence officials warn climate change is a worldwide threat. Inside Climate
        News 2019.

[21]   Vergun D. DOD officials Says U.S. faces climate change crisis. DOD News 2021.

[22]   Hasemyer D. As the world grows hotter, the military grapples with a deadly enemy it can't kill. 2019.

[23]   Refugee and humanitarian assistance. U.S. Department of State.

[24]   National security implications of climate-related risks and changing climate. 2015.

[25]   Facts + Statistics. Insurance Information Institute.

[26]   Extreme weather risks: Restore people's lives. Munich RE.

[27]   Natural catastrophes in times of economic accumulation and climate change. Swiss Re Institute 2020.

[28]   Mills E. Climate change. The greening of insurance. Science 2012; 338(6113): 1424-5.
        [http://dx.doi.org/10.1126/science.1229351] [PMID: 23239720]

[29]   Kottasova I. Climate is the biggest risk to business (and the World). CNN 2019.

[30]   Davenport C. Industry awakens to threat of climate change. New York Times 2014.

[31]   Plumer B. Companies see climate change hitting their bottom lines in the next 5 years. New York Times 2019.

[32]   Flavelle C. Moody's warns cities to address climate change risks or face downgrades. 2017.

[33]   Flavelle C. Lenders' response to climate risk has echoes of subprime crisis. 2019.

[34]   Berardelli J. How climate change is making hurricanes more dangerous. Yale Climate Connections 2019; 8.

Part IV. The Final Verdict

Finally, part IV explains how bad our climate crisis is and what we must do to solve it.

CHAPTER 11

So How Bad is It?

"The clock is really ticking now. It is getting loud."

—Night School, Lee Child

It's not good. We have already increased our global average temperature by 1.8°F and are committed to at least another 1°F.

And the further it goes, the more likely bad things will happen. It's no different than putting on too much weight. A little is okay, but as we go from overweight to obese, the more likely diabetes and other diseases will strike, and the more serious they will be. And now we're at an all-you-can-eat buffet. Our System 2 is telling us no, while System 1 is rushing to the head of the line. That pretty much sums up our addiction to fossil fuels; an assault on our planet for instant indulgence.

Sure, we need energy. We can't live without it, just as we can't live without food. But must it be fossil fuels? Just as we need healthy foods, we also need healthy energy.

We're slowly getting there. But slow and steady will not win this race. Only a sprinting hare will.

CHAPTER CONTENTS

Tipping Points

Bounce on the edge of an Olympic pool diving board. So far, so good. But lean a little too far forward, and you will slip off the edge, and there is nothing you can do to stop your fall. Nothing in heaven or on earth can. You are going to plummet eighty-nine feet into the water. You passed a tipping point.

Global warming also has tipping points. They occur when our environment takes a sudden and unexpected turn for the worse. Once it happens, it can't be reversed. At least not in the next few generations. And that's not the worst of it. It may grab some other tipping points and take them down with it.

Some examples [1]:

1. Polar and Greenland ice disintegration—Melting ice decreases the amount of sunlight reflected back into space, resulting in a spiral of higher temperatures which leads to more ice melting and, therefore, higher temperatures and so on. Like a dog chasing its tail, it never ends until all the ice is gone.

2. Rainforest diebacks—The amount of CO_2 the Amazon forests can store has been steadily declining because trees are being lost from intentional clearing and from fires. As global warming temperatures increase, more trees are lost, so less CO_2 can be captured, so temperatures go higher still. Another unfortunate spiral. Eventually, the rain forests will be unable to regenerate and will turn into grasslands and will never come back. A global temperature rise of 7°F may do it.

3. Permafrost thawing will release methane—The permafrost is melting, and it may release slugs of methane that have the potential to more than double the greenhouse gases now in our air. If it blows, the effect will be catastrophic. A global warming temperature increase of about 9°F will make this more likely.

4. Ocean acidification—As the oceans continue absorbing CO_2, their acidity increases and their oxygen storage goes down, leading to massive fish die-offs, which would put further pressure on an already precarious food supply.

5. The Atlantic Meridional Overturning Circulation (AMOC) is disrupted—The AMOC is a large ocean current that moves northward, transferring tropical heat and moderating the northern hemisphere's climate. But water from melting ice threatens to weaken this flow and could suddenly halt it altogether, leading to extreme weather conditions. For instance, it could seriously reduce Britain's rainfall and destroy its farming industry. A global temperature rise of 7°F makes this more likely.

Tipping points are closing in fast. It could get ugly.

Our Democracy In Peril

Global warming problems could lead to civil strife, creating opportunities for those wishing to sidestep our Constitution to acquire unlawful powers. It has happened numerous times in other countries. Unemployment, reduction in living standards, migration, flooding, disease, food shortages, unequal wealth, national pride, and more can easily destabilize any democracy. Global warming (especially its tipping points) has the power to bring all of this about.

I sense my readers' dismay. *"Do you think we would cave in so easily and abandon the sacrifices our ancestors made for our freedom?"* I don't know, but I've seen glimpses of it already, and let's not fool ourselves by fantasizing we are much different from other populations. We are not. When problems assault our deeply held Metabeliefs, we will grasp at any solution like a drowning man desperate to stay afloat. We will abdicate our voting power to any charismatic leader who promises easy fixes and scapegoats and knows what hot buttons to push. We are not immune, and there is no shortage of leaders eager to jump into the chaos for the sake of power.

The counter is simple. As always, we must turn on System 2 and not make rash, emotional, or uninformed decisions. But the best defense is taking actions that prevent global warming from gaining a foothold. No global warming, no peril. Well, less peril anyway.

Population Correction

It is well established that animal populations increase without bound until their numbers overrun their food supplies. The more of them there are, the more they eat, and the less food is available, leading to a population collapse. Then their prey recovers, and the cycle repeats. In a similar manner, our world population is rapidly increasing though we have managed to stay ahead of it so far. But at the same time, global warming is reducing our resources and throwing other problems at us too. And similar to those seesawing animal population swings, this could be self-correcting for us as well. Global warming's effects could lead to many deaths from disease, starvation, war, and migration.

This is not the outcome anyone wants, so we need to get moving on global warming solutions.

Extinction?

We almost went extinct seventy thousand years ago. DNA evidence shows we hit a bottleneck in which most of humanity died off. No one knows how it happened or how long it took but it almost erased Homo sapiens, preventing us from reaching the twenty-first century. It's hard for me to grasp what it was like for all those people during this Great Dying. People no different from us, identical in appearance and intellect.

And again, no one knows how they managed to survive. But they did. It rested on one or so remaining tribes and maybe only forty breeding couples [2]. The fate of our entire current population of 7 billion was in the hands of just forty couples. Mind-numbing. Good thing they liked each other.

But taking a more cosmic view of extinction possibilities, ever wonder why we haven't found any alien races from the stars? Not for lack of trying. Look at SETI, the Search for Extraterrestrial Intelligence. After thirty-six years and about $90 million, they haven't found a single one. There are certainly some plausible reasons—vastly different technology levels, they don't care to talk to us, we haven't looked in the right places, or maybe we are the only advanced life to have evolved. All possible. But the one I worry about is that there may be disasters that all civilizations encounter when unbridled technical advances outpace the ability to understand and control them. And very few civilizations, if any, can survive these hurdles. So maybe that's why we haven't found any. They could not navigate past one of those choke points.

Here on planet Earth, we already survived at least one: the 1945 nuclear war. Admittedly a bit one sided, but a nuclear war nonetheless.

What other calamities are hiding in the weeds ready to pounce? Global warming? Perhaps. It certainly fits the bill. A degrading environment from greenhouse gas-spewing technologies pushing us to the brink.

So far we've been lucky, but even cats have only nine lives.

The Perfect Storm

Global warming is nipping at our heels. But it is not the only dog chasing us, and maybe not even the most fearsome. There are other calamities quietly waiting their turn, all wanting a piece of us.

Diseases are a good one to start with. COVID-19 is the most recent and most obvious. But what other ones are ready to emerge as we destroy more and more habitats releasing once-dormant viruses?

And what about terrorism, but now amplified by global warming?

Wars? There's plenty to go around and many more are coming our way. A shootout with China or Russia? Quite possible. More strength sapping wars like the Afghanistan one? Getting more and more possible. The United Nations stated that two billion people, that's one-quarter of the world's population, now live in war zones [3]. How about a nuclear war, this time a global one? Not so farfetched anymore.

Mass starvation due to an exploding population and a food supply not equally exploding. It's an old story, and there are plenty of examples in North Korea, Africa, and Syria.

Similarly, our drinkable water is diminishing, which will not keep up with all the new people we keep plopping down on planet Earth.

A supervolcano blowing its top, leading to massive global destruction. Our very own, in Yellowstone Park, erupts every seven hundred thousand years. Guess when the last one occurred? Almost seven hundred thousand years ago. Are we due?

And how could we forget an unseen giant space rock hurtling toward us, maybe at this very moment. It is possible, and heaven knows it has happened in our past. Just ask the dinosaurs.

We might survive any one of these. Not without severe consequences, of course; economies and entire countries may tank. But there may be enough of us left standing to continue humanity's fragile existence. Not the best of scenarios by any means, but a survivable one on the whole.

But what if most of them hit us at once? For instance, what if droughts shattered our food supplies, a super COVID bug started dropping hundreds of thousands of us, Category 5 storms took out New Orleans and the Jersey Shore, flooding made lower Manhattan and the Florida Keys uninhabitable, rising temperatures made a living in the South intolerable, China invaded Taiwan, North Korea set off a nuclear weapon, and so on.

Could we survive such an onslaught? No one can possibly know, but there is some historical evidence for concern.

Around 1177 BC, nine thriving and highly connected civilizations were suddenly and completely destroyed, which Eric Cline describes in his book *1177 B.C.: The Year Civilization Collapsed*[4]. Their cities were leveled with their populations either dead or displaced. And it took centuries afterward for them to recover and regain a foothold and even then, only with a shadow of their former glory.

These civilizations existed in modern Greece and Italy, extending over three thousand miles east to Afghanistan. See map below. They included the Egyptians, Babylonians, Minoans, Hittites, Assyrians, Canaanites, Greeks, and Cypriots, among others.

These were no flash-in-the-pan, here today gone tomorrow civilizations. No, they were entrenched, thriving, and their accomplishments were mind-boggling. They emerged from the Stone Age and founded the Bronze Age with the discovery and commercial fabrication of bronze tools and weapons. They invented the wheel, maybe the most important invention of all time. They also invented the first writing system.

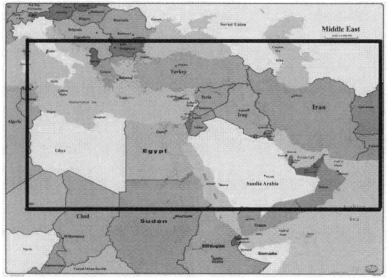

They created the first organized kingdoms with large populations governed by an elite class. They had up to 160,000 people in their cities, whereas before the Bronze Age, villages of 10,000 people or so were the norm [5].

And they lasted for 2,000 years before their collapse. Imagine, 2,000 years and we are struggling just to get past our first 250 years.

Yet despite their longevity and amazing accomplishments, they still fell, and when they fell, they fell hard. Plus, it wasn't just one or two that bit the dust. It was the whole lot of them.

So what happened? Why did they disappear after two millennia of unprecedented success? There were certainly many potential setbacks throughout their existence. Archeologists identified a number of calamities that frequently plagued these people. There were earthquakes, famine, droughts, climate change, insects, crop failures, internal rebellion, and invaders, and their international trade routes were disrupted. As one example, there is evidence of a three-hundred-year drought in Israel, Cyprus, Greece, and Syria.

But as Sanders and Cline point out, these Mediterranean countries have always been hit by earthquakes, famines, droughts, and floods, and they have always survived without a catastrophic, centuries-long collapse.

Cline further states that "none of these . . . would have been cataclysmic enough on their own to bring down even one of these civilizations, let alone all of them."

So why didn't these Bronze Age civilizations recover this time? What was different?

Cline speculates that it was not just a single event that brought them down but the near-simultaneous sum of many catastrophes, which any single one could have easily survived, just as we survived the 2005 Katrina Hurricane. However, each catastrophe magnified the others in a similar way to how tipping points could bring about other tipping points, leading to simultaneous, cataclysmic, and concentrated blows that no group of civilizations, no matter what age they belonged to, could ever withstand.

And if Cline is correct, we may want to take out our worry beads since there are some unnerving parallels between our modern world and those Bronze Age countries.

To begin with, they possessed many of the attributes of our modern world. While they were independent countries, they were also highly connected to one another through economic trade, diplomatic embassies, and social connections. They had many miles of international trading routes in which raw materials and finished products were passed among them. Sort of like Amazon Prime but without the one-day delivery. They exchanged embassies, intermarried, and had considerable social and economic ties. In essence, they achieved complete globalization, in which a butterfly flapping its wings in Afghanistan could cause a hurricane to erupt in Egypt.

On the one hand, this is a good thing since it promotes cooperation and its concurrent advances that no one nation could ever achieve on its own. And it limits the reasons for warfare, though it by no means eliminates them.

But there is a downside to this love feast. One country's problem becomes everyone's problem. Being connected also meant their economies and safety were greatly influenced by each other. As one prospered, so did the others. That's not so bad. But as one suffered, so did the others. And yes, that is bad.

We are exactly the same. We've already witnessed the problems that a fully interconnected civilization can experience when something goes wrong. The COVID-19 virus became a pandemic, killing over six million people and counting, because the world is connected so it could effortlessly hitch a ride from one continent to another. And it disrupted supply chains and severely hurt government economies.

Getting back to bronze in the Bronze Age, bronze was a key economic driver since weapons and tools could be made from it that were much better than anything they had before. But to make bronze, one needs copper, which was in abundant supply throughout the Mediterranean, and tin, which was not. Tin, at least in large deposits, could only be found in Afghanistan. So the export of tin, and the trade routes it took, were critical to the well-being of all the Bronze Age countries. Disrupt the tin supply or its trade routes, and economic chaos followed.

Sound familiar? Oil is our present-day tin. We have fought often over oil, going back to the two world wars, onward to the Gulf War, and now the disruption and soaring prices from Russia's invasion of Ukraine. And untold skirmishes in between.

Now, one would think our technologies would put us at an advantage over the Bronze Age civilizations but I don't think so. Technologies have a sneaky way of wreaking their own brand of havoc and I suspect we are much worse off because of the global warming issues our so-called benign technologies are unleashing upon us. Global warming is also causing droughts, floods, famine, and increased violence all by itself. It doesn't need help from any other catastrophe. It's a one-man wrecking crew.

It is not possible to predict which set of dominos will fall or in what order. But the fact that some will fall is as inevitable as this sentence ending in a period. Thus, the only question of any real importance then becomes, Will they drag us down, or will we be resilient enough to withstand a series of calamities and remain standing even if a bit wobbly? It is not for us to know. All that we can do is work on that which is in our control. Decarbonize. Now.

In the past, I've given our global civilization no better than a fifty-fifty chance of emerging from this century intact, based on any one of the above calamities. And I'm an optimist. But I never considered the combination of many of them hitting us at once. Maybe my optimism was misplaced.

It's Going To Be A Challenge

It's clear to me that we are not going to reach the IPCC target of being greenhouse gas-free by 2050. Some disagree, and I hope they are right, but there are others who are on my page. In 2015, a study stated that keeping global warming to 2.7°F (1.5°C) was probably not possible and that an overshoot was going to happen [6].

A climate researcher echoes this sentiment and lays the blame on overly optimistic scientists giving the public false hope. He is convinced that even a 3.6°F goal is not attainable, and he doesn't think recovering from the overshoot is possible with the technologies in hand.

A report for the UK government stated that achieving zero emissions by 2050 "exceeds even our most speculative measures . . . that push against the bounds of plausibility" [7].

Some reports say we only need a political will because the technologies exist, and it will just take a pittance of our GNP to solve global warming. These reports are seriously flawed and, like bad medical advice, only make matters worse. While some technologies certainly do exist, solar and wind power, for instance, other critical ones do not. Making nuclear fission reactors safer, cheaper, and widely deployed is one of those, and getting fusion reactors anywhere near commercial operation is yet another. The misunderstanding is that some think we can get by with just a fix here and another there, but the truth is we are so far behind the eight ball that we will need all the tools we have and a few we don't, to get us out of the jam.

Then there's that old nemesis of ours, time. We have less than thirty years to get all these technologies developed, commercialized, installed, and to get most of the U.S. electrified. This is a war-footing type of change with a war-footing type of sacrifice that people are not going to agree to. Perhaps if a worldwide global warming calamity occurred, then that would shake people up, but isn't that what we are trying to avoid?

We have already been making impressive progress toward zero emissions. But achieving it in the limited time left? It's not happening. Across the U.S.? No. Across the whole planet? Not a snowball's chance in a global-warming-heated hell.

And We are Dragging Our Feet

The problem is we don't view global warming as a near-term problem so we put it off to save our efforts for more pressing issues. Makes sense. After all, as we evolved, it was those who could survive short-term dangers who prospered and passed on their genes. The need to deal with long-term threats, while certainly not unimportant, was less of a force in determining which genes moved forward.

Before the Industrial Revolution, the average lifespan was only twenty-five to thirty-five years [8], so there was little profit in worrying about distant threats. Most were not going to be around to tackle them. That was a good way to build humans, but now life has become more complex and we live much longer than ever before.

Global warming, while not seen as an immediate threat, is an unrelenting one inexorably marching toward us. And, we can only effectively deal with it piecemeal, one ponderous step at a time, over decades to make small improvements that will add up over the long haul.

It's like a massive planet-killing asteroid hurtling toward Earth. It does no good to wait until the last moment, right when it's about to strike, despite Bruce Willis's heroics in *Armageddon*. Life doesn't work like that. We need to start nudging the asteroid as soon as possible so that there is time for continuous gentle pushes, over years, to move it from our path.

The same with global warming. And time is running out. We need to start now not later. But those pesky genes keep getting in the way, sabotaging our efforts.

That doesn't mean if we don't solve global warming within the next thirty years, we will be greenhouse-gassed back into the Stone Age. No, though we better make some substantial progress or we will be heading in that direction. The COVID disruptions will be a walk in the park compared to this.

But We are Making Progress

Our greenhouse gas emissions have not increased since 1990[i] and renewable energy prices have declined, leading to a doubling of their use since 2008. We have cut back on burning coal by 40 percent since 2007, which is the worst of our greenhouse gas offenders. Nuclear energy continues to make strides. And many more people are admitting global warming's consequences into their Metabeliefs, begrudgingly maybe, but within the fold nonetheless. So at least we are not bringing a knife to a gunfight. No, we are better off than that. But we need to continue and to accelerate our actions.

The clock is ticking.

Connecting the Dots

1. Tipping points are waiting in the shadows, ready to pounce. And once they strike, there is no turning back. Ice disintegration, deforestation, methane release from thawing permafrost, ocean acidification, and ocean current disruptions are a few to worry about.

2. Global warming problems may put our democracy in peril. As good a job as our founding fathers did in creating our government's checks and balances, nothing can ever be foolproof, especially when our Metabeliefs are attacked. We will willingly ditch our democracy and embrace any dictator who promises a return to the good old days, however illusory they might be. Global warming may be the final straw that creates such a situation.

3. Human extinction? It could happen. Running through the technology-created minefield of ours is dangerous enough, but now we have added global warming issues to boot.

4. But it's not just global warming. We should be so lucky. It is also about a host of other problems running neck-and-neck with global warming that is coming at us like a cattle stampede.

5. And time is running out. We have less than thirty years to decarbonize or face significant global warming calamities.

6. Yet despite our Metabeliefs and genes and what have you working against us, we have been making progress. Who says you can't teach old human new tricks? But we must continue and accelerate.

Afterthought: God's New Covenant

In a one-tablet announcement, God declared that He is canceling the Animal Kingdom and will instead focus His attention on rocks. "I like rocks," He is quoted to have said. "I've given humanity the most beautiful planet in the galaxy with wonders poets never dreamed of, and all I've asked in return is they take care of it. Such a small thing, really. Take care of that which gives you life and sustenance. Could I have made it any easier or any clearer? Yet, in mind-bending arrogance, they've managed to choke the life out of it and stubbornly refuse to repent.

So it's time to give rocks a try."

When asked in a follow-up press conference what actions would be taken, He responded, "None needed. Humanity is taking care of it for me. At least that's one thing they've gotten right."

---

[i] Most of the world's greenhouse gas emissions have risen little since 1990, except for China, which has been on an all-you-can-emit binge.

References

[1]   McSweeney R. Explainer: Nine tipping points that could be triggered by climate change. Carbon Brief 2020.

[2]   Krulwich R. How human beings almost vanished in 70,000 B.C. 2012.

[3]   Archie A. World is seeing the greatest number of conflicts since the end of WWII, U.N. Says. NPR 2022.

[4]   Cline EH. 1177 BC The Year Civilization Collapsed. Princeton University Press 2014. [http://dx.doi.org/10.1515/9781400849987]

[5]   Historical urban community sizes. Available From: https://en.wikipedia.org/wiki/ Historical_urban_community_sizes/

[6]   Mooney C. Scientists: This is what it would take to keep the world super safe from global warming. 2015.

[7]   Harrabin R. Climate change: UK 'can't go carbon neutral before 2050. 2020.

[8]   Hodges P. Rising life expectancy allowed industrial revolution to occur. 2015.

CHAPTER 12

Can Anything Be Done, or is It Game Over?

"The difficulty is to persuade the human race to acquiesce in its own survival."

—Bertrand Russell

A doctor prescribes meds to break a life-threatening fever. I think we would gladly and thankfully take the pills.

This is pretty much what the IPCC has told us [1]. Our planet has a fever, and it has already warmed by 1°C (1.8°F)[i] due to global warming and will climb by another ½°C (1°F) somewhere between 2030 and 2050 if we don't take action. And the consequences could be dire, with monster storms, droughts, food shortages, increased violence, and much more. So to avoid the worst, we must completely eliminate our greenhouse gases by 2050.

The medicine we need takes two different approaches, both necessary.

The first is obvious. We need to install equipment and take other actions that will reduce and eliminate greenhouse gases to stop this relentless global temperature rise. Some are ready to go, some need development, and still, others need considerable development.

But not just any equipment. No, there is an unfortunate rub to this. The IPCC has put a hard stop at the year 2050 to decarbonize our planet so as to avoid the worst of what global warming has to throw at us. And this deadline must factor into our decisions. We need to drive the technologies that are more likely to be widely deployed by then. That's not to say we should avoid longer-term ones, but rather we should be putting most of our efforts and resources into those that can get to us before the 2050 window closes.

The IPCC deadline also means we need to get serious now not later. Our way of living has always been, "Why do something today if I can put it off till tomorrow?" That must cease. It must be "Do it today, do it right, and do it quickly."

And now is not the time to pick winners based on political or personal agendas or financial gain, which, unfortunately, I have seen. For instance, there's the renewables camp which dismisses nuclear as having too many disadvantages. Then there are the nuclear proponents who claim renewables can never get us far enough. I've run across the carbon tax zealots who believe it to be a new religion we must all bow down to. Each thinks theirs, and theirs alone, must be pursued. We cannot pick a single approach. Risky business that. We need them all, and at this point we can't predict which will work well enough to be part of the mix. Some are not going to work. That's okay. Discard them. Move on.

The second path is less obvious but just as important. We need to take control of the problem as individuals, and as part of the groups we belong to and be sure those representing us do the right thing. We need strong leadership, and we must give them our complete support to solve an ominous civilization-wide problem. And if they don't, then we must replace them. We must cut through the foolishness that so often accompanies emotional issues. That is squarely on our shoulders and to fail at this is to fail our nation and our children.

We are not going to make it all the way to the IPCC goal by 2050; there's just not enough time, resources, and will. But we must get as close to it as we can so we can pull ourselves back from the cliff's edge when the technology and social issues catch up.

It is not going to be easy. But it is possible and it is very much necessary.

---

[i] Compared to the average temperature measured between 1850 and 1900.

CHAPTER CONTENTS

Technologies We Must Install

Renewables Work and Should Be Expanded

Imagine a city run entirely on renewables. What would we hear? What would we smell? There would be no pollution from fossil fuel burning, so maybe we would smell flowers and the aroma from bakeries and cafes. There would be no noise from automobile engines, so maybe we would hear each other talking and music from local shops. What a thought this is. Imagine the tranquility of not having our senses under constant assault but rather massaged by what is natural and pleasing. And the peace of mind that we are finally defeating an enemy threatening worldwide calamity. Imagine all of this. This is what renewables bring. All our energy needs, from automobiles to heating our houses to running our factories to growing our food, would be based on electric energy. And that electric energy would come from the wind, the sun, tides, and waterfalls. Clean, pristine energy. And it is already here and growing. This is worth imagining.

Renewables are energy sources that produce little or no net greenhouse gas emissions either by their nature or because they can be replenished by soaking up atmospheric CO_2 equal to what they generate. Solar, wind, hydro, geothermal, and tidal are examples of inherent greenhouse gas-free energy sources. Renewables are an important part of the global warming solution since they work and are already being used.

A crucial aspect of renewables is that they are well suited to replace fossil fuels in electric power generation, where most of our energy will need to come from. However, they will have a more difficult time in other arenas, such as transportation, and heavy industry, and certainly are not likely to be used much in air transport.

For renewables to have a large impact, we need to convert all fossilfuel–fired devices into electric ones. We are making modest progress. Electric cars in the U.S. have increased from 224,000 in 2018 to 1.2 million in 2019 [2]. Admittedly, electric cars only make up less than 2 percent of the total, but given all the hype and money being put into it, this will likely increase.

In the U.S., renewable electric generation has gone from 9 percent to 18 percent since 2008 [3]. Google, Amazon, Microsoft, and Apple are driving this increase due to their commitment to feed their power-hungry data centers with renewable energy.

Fortunately, renewable costs have been rapidly declining. Globally, electric generation from onshore wind and solar now costs about the same as fossil fuel plants [4]. Plus, the installed cost of solar is in the range of fossil-fired plants and much cheaper than nuclear or fuel cells [5]. Further cost declines are expected.

Some energy sources are not traditionally considered renewable, but they should when viewed through the global warming lens.

Hydrogen fuel is one of them. You obtain hydrogen by separating it from oxygen atoms in water, and when you burn hydrogen, it goes back into the water.[ii] Can't get more renewable than that. Plus, it produces no greenhouse gases, and you get oxygen as a byproduct, so even better. But the problem is that you need the energy to separate the hydrogen from the water and, worse still, you expend more energy than it provides. So the only way hydrogen works as a fuel are if you use another renewable energy source, such as solar, to produce it. That means you are only converting one renewable for another, with a commensurate loss that entropy always demands, so why do it?

There are a few reasons. It is an ideal way to store renewable energy as it can be transported in pipes, of which an extensive network exists to carry natural gas. Plus, it may be able to replace fossil fuels used in long-haul shipping and airplanes that could not possibly be powered by renewables or nuclear energy.

And it can replace fossil fuels used in the cement and steel industries that cannot easily electrify.

The European Union expects 50 gigawatts of hydrogen energy production by 2050, which is the same as fifty nuclear power plants [6].

Recycling is another one of those energy sources not considered renewable, but it has renewable traits because it recycles the energy that was originally used to make it originally. Take an old beer or soda can. It certainly takes energy to recycle them, but it only takes 5 percent of the energy needed to make a new can from scratch. Hurrah for us. But only if we recycle. And for every ton of cans we recycle, we eliminate 10 tons of CO_2 emissions [7]. Not bad at all.

"As an Aside, here's a cocktail party question: How many times can you recycle the same can? Forever? Guess again. One thousand times? Not even close. "Okay, tell me." I will. Entropy, in its inexorable march to oblivion, has its sticky little fingers in everything, and cans are no exception. To recycle a can you need to melt it, form it into a sheet, then roll that sheet into a new can. But at every step of the way something will always go wrong. Not much, but something. For this can, some of it will combine with oxygen, sending it back to its original ore. So maybe we will recycle it twenty times or so. No more. And that's not including cans that never make it to the recycling bins at all. Nonetheless, recycling is

better than not, so let's take in as many as we can (Can. Get it?)".

Renewables, by themselves, are not the complete solution. But they don't have to be. They just need to buy us some time and combine with other solutions to bring us home.

A Carbon Tax Adds to the Mix of Tools We Have

When we visit our doctors, part of our copay goes to cover their malpractice insurance. The cost to settle the lawsuits from their medical negligence is included in their fees. This is true for almost any product we buy, which has built into its price the cost for warranty claims and repairs. But that's not the case with fossil fuels. The true cost of burning fossil fuels goes beyond the price we pay at the gas pump, which includes only the cost to make and deliver the energy to us.

But a much larger penalty is the damage it causes by creating CO_2, which warms our planet. Not all fuels are equal in this regard. Coal is the worst offender, natural gas so-so, and renewables and nuclear energy not at all. So it makes sense to account for these costs the same way your physician does.

A carbon tax is an easy way to do it. It is a fee added to the cost of the fuel that is proportional to the carbon it contains. The tax is directly related to the amount of CO_2 it produces and therefore to the damage it causes. Fuels such as coal will have a high carbon tax, and renewables and nuclear none at all. Plus a carbon tax, if properly done, levels the playing field so we can compare nuclear and renewables to fossil fuels.

However, that's not the only reason to impose a carbon tax. Rather it is imposed because it reduces fossil fuel use, thereby reducing global warming. Why does it work? Simple economics. The more something costs, the less it will be purchased. During the 1970s oil crisis, gas prices were pushed to record levels. As a result, people stopped buying gas-guzzling SUVs and instead bought smaller cars. What happened when gasoline prices dropped? SUV sales skyrocketed.

What should the carbon tax be? There is no good answer to this, as there is no way of knowing how markets will react and what the dynamics are among the other possible greenhouse gas-reducing techniques available. So we take our best guess and revise it as time goes on and more is learned. No real big mystery here, as we will need to experiment to determine the optimum number. And it will change over time as the amount of emitted CO_2 goes down.

We have a starting point from various studies that are not too far apart [8]. A tax from \$15 to \$55 per ton of carbon will reduce CO_2 emissions 6 percent to 50

[ii] It can also be obtained by gasifying wastes containing hydrogen such as municipal solid wastes, from natural gas (which contains hydrogen) and from any other fuels containing hydrogen.

percent by 2025. This will increase the cost of a gallon of gas by 4 cents to 15 cents. Not insignificant but also not an overwhelming burden.

What do we do with the revenue? That's for the voters to decide, but giving it back to individuals via tax refunds would pose no direct burden on taxpayers, which is another advantage to a carbon tax. At least in theory.

Similar to a carbon tax, a cap and trade is another way to cover extra costs. Instead of taxing the carbon, it puts a limit on how much carbon can be emitted and lets low emitters sell credits to the high emitters to offset their emissions. The market then decides on the carbon credit price.

Gilbert Metcalf explains the difference between the two [9]. A carbon tax puts a price on CO_2 and lets the market determine how much to pollute, whereas a cap and trade puts a cap on pollution and lets the market determine its price. An advantage of the cap and trade over a straight carbon tax is that it turns carbon capture into a revenue-generating business which would, therefore, not rely on taxpayers' largesse to pay for it. Anytime a company uses carbon capture, they can sell the carbon credits to another company, which could offset the real costs of carbon capture and perhaps make it work as a standalone, profitable business.

Like everything else in life, a carbon tax has problems all it's own. First, there are always ways of getting around any legislation, which brings to mind Ambrose Bierce's definition of a lawyer as "one skilled in circumvention of the law." This needs to be prevented. All major emitting countries need to sign up for it and must be held accountable; otherwise, it puts complying nations at a significant cost disadvantage.

A second major difficulty of using a carbon tax is that it taxes fuels based only on the CO_2 released as they burn, which completely misses the problem of natural gas leakage, a major source of greenhouse gas emissions.

Third, are we really sure the tax revenue generated will get back to the consumers? I am not. The good intentions of the federal or state legislatures that enact this tax don't matter, because an important attribute of any democratic legislative process is that a current body cannot bind the wishes of a future one. So a future legislature could easily appropriate the tax revenue for other purposes.

Still, despite the problems, a carbon tax is a valuable tool to consider along with the other techniques available to us. Plus, several oil and gas companies believe it would reduce uncertainty and stimulate low-carbon investments [10].

Nuclear Energy—The Comeback Kid

Remember those molecules we played with in chapter 4? Here's another trick we can do with them. Take a uranium atom and split it in half in a process called *fission*. Not so easy to do, but if you can, then neutrons will be emitted and a bit of the atomic mass will disappear and turn into energy. The neutrons will crash into other uranium atoms, which will split more atoms and release more energy and more neutrons, which then do the same in a continuous cascade. This causes a sustained nuclear reaction, continuously giving off energy in the form of heat and radioactive emissions.

Once you have the energy generated from the nuclear reactor, it's easy to produce electricity using technology dating back to the 1700s, when the first industrial steam engines were built. Use the heat from the reactor to heat up water, which changes it to steam, and in so doing, increases both its temperature and pressure. Take the high-pressure steam and use it to turn a turbine, in a similar fashion that whistling steam coming out of a teapot can turn a plastic paddle placed in its path. Connect the turbine to an electric generator that then produces electricity which you finally move through copper and aluminum wires to homes and factories.

In the U.S., we have ninety-six nuclear reactors in fifty-eight plants producing 19 percent of our electricity [11]. A nuclear plant will provide the electric needs for almost half a million homes.

Some studies have shown that nuclear energy is safer than making electricity with either coal or natural gas. One study's calculations state that for every death resulting from nuclear energy, you would get 3,300 deaths from coal-fired power plants and 280 deaths from natural gas–fired power plants for the same amount of energy production.

I'm not so sure. While I have no doubt the study is correct as far as it goes, it missed some serious issues. Along with the fission process of splitting the atoms come significant byproducts of radioactive waste that need to be stored for millions of years. Yes, you read that right, *millions* of years. And this radioactive waste is deadly. You don't just throw it into your neighborhood landfill. It needs to be shielded by storing it in drums of steel and concrete, so its radiation doesn't leak out and harm anyone and it needs to be cooled, typically with water, since it will always be giving off heat.

When you shut down a gas-fired or coal-fired power plant, you shut it down forever with very few residual obligations. But when you shut down a nuclear plant, its ghost never goes away. And every time you build another one, it is a legacy that keeps on giving and lasts hundreds of thousands of years or more.

As of 2017, worldwide, 441 operating reactors generated 11,000 tons of highly radioactive spent fuel every year, with 210,000 tons currently in storage [12]. Assuming we want nuclear energy to produce 50 percent of our electricity, we would need to increase that to 50,000 tons of radioactive waste each year. After ten years, we would have accumulated 710,000 tons of waste. After one hundred years, it would be 5.2 million tons of high-level radioactive waste. Not good. And not too comforting as it lasts forever. I say forever because our kids, our grandkids, and then another 33,000 generations after them will need to live and cope with this ever-expanding pile of death. And the more we produce, the more generations will have to deal with it.

And never mind that terrorists are not likely to target a coal-fired power plant but instead would gladly give up everything dear to them to successfully attack a nuclear plant. I bet the researchers didn't have that in their calculations.

The real issue is that nuclear has some problems that cannot be quantified with the same accuracy as the other energy sources. It is not so much that this makes nuclear more dangerous, but rather it makes it more uncertain, so comparing fossil fuels to nuclear is comparing apples to oranges. On the fossil energy side, the dangers are larger and reasonably well-known. With nuclear, the dangers that can be thought about are not so bad, but the dangers that cannot be analyzed create an unending future menace

And if that isn't enough, I was listening to a nuclear industry executive right after the Fukushima nuclear accident in Japan. In true doublespeak fashion she said that the Fukushima accident proved that U.S. nuclear plants are safe. So an accident proves it is safe. That was ten years ago and my head is still spinning from that logic. Does that mean I should crash my car to prove to my insurance company that I'm a safe driver?

The technology problems are bad enough, but sometimes the people in charge of the technology are what really scare the bejesus out of me, especially when there is a lot of money involved. And this is where all the studies in the world completely miss the mark.

Still, if you can ignore these ageless elephants in the room, nuclear has some safety advantages that the public is generally unaware of. The chief one is that nuclear energy will avoid some of the deaths arising from global warming. According to the World Health Organization in 2000, there were 150,000 deaths directly attributable to global warming, and they conservatively estimate that by 2030, it will rise to 250,000 deaths [13]. I don't know how accurate this is, but whatever the true numbers are, it will only get worse, maybe much worse, as time proceeds and global warming effects accelerate. The use of nuclear energy would certainly resist and perhaps even reverse this trend.

But there is another, more practical and urgent tradeoff to be made. Global warming problems are here and now, are getting worse, and will not end if we don't intervene immediately. In fact, global warming issues will lead to serious worldwide calamities if not acted on at once. WHO has already warned us about the deaths directly attributed to global warming. Nuclear is one of the solutions that can slow this down and eventually reverse it. Whereas the issues with nuclear energy, serious as they are, make no mistake about that, occur much further into the future and can be dealt with later. Besides, if we don't solve global warming first, we will not be around to worry about nuclear's problems anyway.

Consequently, we must pursue the nuclear option despite its warts. The simple fact is nuclear produces copious amounts of energy 24/7 in a highly compact form, and most importantly, with no greenhouse gas emissions.[iii] No other energy source accomplishes this; *none*, which is an overwhelming point in its favor.

Much work is needed, as only one nuclear plant has been built in the U.S. in the past twenty years, five have retired in the past five years, and twelve reactors at nine plants have announced plans to retire ahead of schedule [14]: "Globally, nuclear power has declined from 17.6% in 1996 to 11% in 2019 [15]".

Also, nuclear energy is much more expensive than fossil-fired power plants since natural gas is at a historically low price and nuclear's zero greenhouse gas emissions are ignored, putting it at an unfair cost disadvantage. Interestingly, a carbon tax would fix this drawback to some degree: "A $15-per-ton carbon price would hold U.S. nuclear capacity roughly steady through 2050. A $25-per-ton carbon price would substantially increase capacity" [16].

A lot has been learned since the Chernobyl and Three Mile Island accidents, for sure. There are some interesting designs being worked on that could be cheaper and safer than current designs. One of them, small modular reactors, or SMR, can be built in factories rather than the more expensive way of building on-site, and they use off-the-shelf components, both of which greatly reduce their cost. NuScale Power is an example of an SMR manufacturer that is trying to make units from 50 to 200 megawatts with a prototype cost of $100 million. Based on these numbers, that's an installed cost of $500 to $2,000 per kilowatt [17], which compares favorably to fossil-fired plants at around $1,000 per kilowatt.

Plus, new designs called Generation IV systems use alternate cooling, and one of these designs uses spent nuclear fuel, depleted uranium, or uranium in which no processing is required [18].

Nuclear fission has many advantages going for it. It works, it has been in commercial use for over sixty years, it does not produce any greenhouse gases, and there is considerable ongoing research to improve it further. Given what is at stake with global warming, we would be foolhardy not to pursue it.

"As an aside, what about launching the radioactive waste into space? Pie-in-the-sky thinking I know, but if we could, it would eliminate nuclear energy's one and only drawback. We would need to launch the waste fast enough to escape Earth's orbit, with absolutely no chance of an accident during the launch that would rain nuclear debris over its flight path, and do it cheaply and without any net greenhouse gas emissions".

"The technology doesn't currently exist. But who knows what technologies will be available in the next one hundred years, or even longer. At least nuclear waste's one disadvantage, that it lasts forever, plays in our favor this time. It gives us plenty of time to mull it over".

Is Fusion Energy the Holy Grail That May Never Be?

Do you want nuclear energy without the bothersome baggage of millions of years of radioactive waste? If yes, then you've landed in the right section of this chapter. Nuclear *fusion* produces energy similar to nuclear fission from the previous section, in that it harnesses the energy locked up in atoms.[iv] But it greatly differs from fission by fusing two light atoms together rather than splitting one heavy one apart, thereby avoiding the nasty radioactive waste issue that plagues conventional nuclear fusion plants. Two isotopes of hydrogen atoms (deuterium and tritium) are smashed together to form a helium atom, and in this fusion process some mass is lost, which turns into energy.[v]

Fission involves tearing an atom apart, and from your own experience tearing apart anything, say a loaf of bread, will leave unwanted crumbs. Fusion, on the other hand, smashes two things together, such as two blobs of Silly Putty, which will merge smoothly with much less left over. Fusion's nuclear waste products are nothing more than radioactive tritium which has a half-life of decades, not millions of years as in nuclear fission.

Both deuterium and tritium can be easily extracted from seawater and there is enough to power the world for six million years [19]. As a further comparison, it would take 10 million pounds of coal to yield as much energy as one pound of fusion fuel [20].

[iii] There will be some greenhouse gas emissions from constructing the plant, decommissioning it, and refining the uranium. But the amounts are small compared to the emissions of fossil-fired energy.

Nuclear fusion provides unlimited energy using cheap seawater as its fuel, produces no greenhouse gases, and little nuclear waste, which is easily handled. So what's not to like?

Plenty, which I'm sure you saw coming. After sixty years of research and $50 billion invested [21], and a current annual budget of $900 million, fusion has yet to produce an ounce of net energy. Nothing, zero, nada. That's like having a stable of race horses and after sixty years of trying, not one has ever finished a race, never mind winning one. Not much good and not anything to bet on.

The problem is that for nuclear fusion to work, it is necessary to duplicate the condition of the sun's interior right here on Earth. The reason? With fusion you need to have two atoms merge together, but they each have a positive charge from the protons in their nucleus, which will repel each other the way two magnets of the same polarity repel. To overcome this, the atoms must move very fast, so fast in fact, that they must achieve a temperature of 180 million degrees Fahrenheit, which is six times hotter than the sun's core.

Which leads to not just any problem but a monumental one. When you boil water for your morning coffee, you need a pot to hold it in. No different with this plasma, but what will hold a mini-me sun? No material could ever contain this unimaginable fury of temperature and energy that would blast anything coming close to it into tiny subatomic bits. So another tack is taken, magnetic or laser confinement. Because the fusion gas is so hot, it becomes a plasma with charged particles. That is, electrons are stripped off the atoms so the remaining nucleus becomes positively charged. What better way to control a hot charged gas? Use a magnet that can contain the gas into a fixed volume without ever having to touch it. The magnetic lines of force radiating out from the magnet do the dirty work and they aren't bothered by the high temperatures. Lasers can perform the same task, and some designs use a combination of the two, both magnets and lasers.

As difficult as that sounds, researchers have accomplished this many times. I know that's not something you want to hear, that some wild-eyed scientist is making mini-me suns just down the block from you. Not to worry. They don't last very long and they are quite tiny. No, not the wild-eyed scientists. The mini-me suns.

But the energy to create them and then to control them is far larger than the energy fusion gives off. Despite these exceptional achievements, no net energy has ever been produced. Until that occurs, all bets are off, and even when it does occur, all bets are going to be long shots.

[iv] "Fusion" and "fission" sound similar but are quite different and should not be confused with one another.

[v] Isotopes have the same number of protons but a different number of protons in the nucleus. Their weights are different but their chemical properties are the same.

Some researchers therefore claim that the production of net energy, even if it's just a tiny drop, is the Holy Grail of nuclear fusion and will whisk us away to unlimited nearly free energy forever. And they will break open the champagne, assemble the press conferences, go on the talk shows, get their pictures in the scientific journals, and, all in all, will be happy and content with themselves.

All of which is an illusion, a very sad illusion. We might as well have saved all those billions of dollars if that's all we get out of this. No, the prize that will do humanity any good at all is much further off than that, and equally as difficult. That ounce of net energy needs to be turned into a tsunami of net energy at a cost that is competitive with current energy prices. The fusion energy plant must be built for something less than a king's ransom and it must work with little maintenance, operating 24/7 without interruption. I can tell you from my own experience with large and complicated technologies, getting to net energy is the easy part. Getting it to the point of commercial success is where the real work begins. I could fill volumes with brilliant ideas I have worked on that never got past the lab stage. And nuclear fusion sits at the pinnacle of complications and difficulties. Putting two men on the moon in 1969 was child's play compared to turning fusion into a commercially operating plant.

Which would be okay if we could wait for it, except that time is not kind to us. The big question is not when will fusion produce net energy, but when will it produce it commercially and at a large enough scale that it would replace a large portion of our fossil fuels. That is the $64,000 question, although perhaps more appropriately the billion-dollar question. Fusion must be working as an everyday garden-variety energy plant in the next 20 to 30 years at most, for it to help us at all. And that is simply not going to happen because of the enormous technical difficulties of lassoing and controlling a mini-me sun and then the formidable commercial issues of siting, permitting, building, and operating a fusion plant.

Nevertheless, while fusion will not be the cavalry arriving just in the nick of time, it serves another vital niche in the global warming battle plan. It tag-teams with fission as follows. Fission is necessary and it exists and it works so there are no technical uncertainties. It will fill a void we desperately need to decarbonize our energy production. But while nuclear fission helps in solving our near-term and serious global warming problems, it silently creates a problem almost as bad as its radioactive waste.

Therefore, my suggestion is to rapidly ramp up nuclear fission while we wait for nuclear fusion to become a commercial reality. Once it does, fusion can step in and replace fission with the same advantages but without the radioactive waste legacy. Then, depending on how long it takes for fusion to mature, we will have a limited pile of fission waste to deal with and it will no longer accumulate. This I can deal with. We sort of get the best of both technologies and use one to offset the other's disadvantages. Time, while no friend, becomes less of an enemy and more manageable. Fission addresses our near-term issues and fusion solves fission's far-term issues. And if we choose poorly and fusion doesn't work, we are no worse off (though a few billion dollars poorer). So what's not to like?

Methane Leaks and Cow Belches: The Forgotten Greenhouse Gas

Methane lasts for only decades, so it is not as bad as CO_2, which lasts for centuries. Nonetheless, methane definitely affects global warming, as pound-for-pound it is a much more potent greenhouse gas than CO_2 ever could be. Overall, it makes up about 10 percent of our global warming problems. So it should be targeted.

And targeting it is actually pretty straightforward, unlike many of the other solutions we need to go for. Methane gets into the air and gives us agita primarily in two ways—natural gas leaks and cow burps.

Natural gas is mostly methane, and it gets around by being forced through two million miles of pipe, so it can be delivered to our homes for heating and cooking and to industry for multiple uses. Any idea what 2 million miles of pipes look like? That's enough pipe to wrap completely around the earth. Then again. And again. And seventy-seven more times over. Now, that's a lot of pipe.

But some of those 2 million miles of pipes are quite old and will leak the way an old garden hose springs small jets of water. In fact, in the U.S., the oil and gas industry emits 14 million tons of leaked methane each year and the world emits about 83 million tons [22].

The good news is that leaks can be easily fixed. Maybe not with duct tape (though I wonder), but the leaks are easily found and really all it takes is money to replace or repair the leaking pipes. So what it really comes down to is the amount of money saved by stopping the leaks compared to the cost of fixing the leaks. Certainly, if the environmental cost is added to the value of the methane, it could well be worth an investment that industry would be willing, if not eager, to make.

And what about those burping cows? All 100 million of them. We have more cows than most countries have people. In fact, if our cows went rogue and formed their own country, it would be the fifteenth-most populous one in the world, well ahead of every European country. By a lot. I wonder what their country motto would be? *Eat more chicken*? Their national hero? *Colonel Sanders*? Their favorite sport? *Cow tipping*? Okay, I realize this is getting utterly silly, so let's moo on.

Cows contribute 28 percent of all methane emissions in the U.S [23]. And just like the eighty-times-around-the-world gas pipe leaks, cow burps can also be easily fixed. Simple. As their own motto states, eat more chicken and more fish and more vegetables and more fruits. Plus, our food costs would go down and more land would be available, as the pastures would no longer be needed. And our health would improve. A triple bonus. Everyone wins. Well almost. Not the cows. But hell, those proud cow insurrectionists wanted to secede, so this is the price they pay. Democracy lives on.

Carbon Capture Has Now Become Essential. We Can't Leave Home Without It

We are not going to achieve the greenhouse gas reductions required to stop global temperatures exceeding a 1.5°C rise by 2050. With less than thirty years remaining and the political climate and lack of concern being what it is, there is no credible path that I can see. And depending on how poorly we do, it could be disastrous for us beyond any previous calamity we have faced. And once we overshoot the 1.5°C, because of the cumulative amount of CO_2 in the air, there is literally no way back since CO_2 remains in the air forever.

So all options must remain on the table, and carbon capture is an important one, especially after our inevitable overshoot. There are two steps to this: (1) remove the CO_2 and then (2) store it somewhere so that it does not reenter the air. And there are three ways of achieving these two steps.

The first, and technically the easiest, is to both plant trees and stop forest destruction. Plants soak up CO_2 as they grow and hold onto it for maybe a century before they die and release it. This is the best natural sink for removing CO_2 we have. And let's take this a step further. We can reduce the grazing land and turn that into forests if we eat less beef.

The second way is to pull the CO_2 out of the flue gases being emitted by electric-generating power plants when they burn fossil fuels. The exhaust gases from these plants are bubbled through water containing amines, an ammonia-like compound, in which the CO_2 molecules attach to an amine molecule, and then the water is heated to release just the CO_2 amine combination. Next the CO_2 gas is pumped down deep into the ground or ocean where it cannot escape.

The third way is forcibly removing CO_2 from the air around us using direct air capture technologies, and then to again store the CO_2 underground or in the sea.

Each option has advantages. The forest one is the least expensive and there are no technical risks. It works. Flue gas capture has the most concentrated CO_2 of the three so it is the easiest to remove (apart from the flue gases high temperature). For instance, coal-fired power plants have CO_2 concentrations between 17 and 83 percent, whereas, in the air, the CO_2 concentration is around 0.04 percent. Finally, direct air capture can be done anywhere in the world, so siting the equipment opens up many options.

Unfortunately, carbon capture is not cheap. James Hansen, quoting a study by David Keith, says the cost is $102 to $208 per ton of CO_2, including storage [24]. Another study predicts that CO_2 capture could be down to $90 per ton by 2025 if storage costs are included [25]. They further claim a 13 percent reduction by 2050. Yet, another study claims that by 2025 it will cost $20 per ton to remove CO_2[26].

So I've gone from $208 to $90 to $20 from a quick search. Maybe if I keep digging through the literature, the cost will become zero or even less. We really need to take technology projections with a grain of salt and maybe a stiff drink too. So I am going with Hansen's numbers. At least his are based on hard experiments with no projections made. That's not to say that this will scale, but let's go with the best info we have.

Globally we emit 40 billion tons of CO_2 per year. So that would cost (at $208 per ton) $8 trillion per year to capture it all. The U.S. GNP is $22 trillion and the world's is $80 trillion. That's more than a third of the U.S. GNP or 10 percent of the world.

Direct air capture may be able to remove 10 to 15 billion tons of CO_2 per year by the end of the century, while others think it could be as high as 35 to 40 billion tons. Sabine Fuss disagrees and thinks at best, we could remove only 5 billion tons at the cost of $100 to $300 per ton [27]. That would be $5 trillion per year.

Reclaiming forests and planting trees can soak up to 3.6 billion tons of CO_2 per year, at the cost of $5 to $50/ton. Better, but still expensive, from $18 billion to $180 billion annually.

Of course, these costs should be compared to the damage global warming will cause worldwide, which is $2 trillion to $10 trillion per year [28].

There is ongoing research on CO_2 capture. According to the U.S. Department of Energy's National Energy Technology Laboratory, as of 2014, there were seven carbon capture demonstrations operating, with another twelve in evaluation or design phases. Additionally, throughout the world, there were six operating and twenty-one in evaluation or design phases.

And Resources of the Future stated: as of 2018, there were twenty-three large-scale carbon capture and storage projects underway, capturing about 44 million tons of CO_2 annually [29]. True, that's not a lot, but more importantly, they are showing it can be done and it works.

But there are some serious problems with carbon capture that must be overcome. The first and obvious one is the cost. We are never going to deploy carbon capture unless the cost gets down to the single digits per million tons removed and it will probably need to get down even lower.

A second problem is scaling up the research demonstrations to real-life commercially operating plants, which is always a serious issue anytime new technologies are being developed. An example illustrates this. Imagine you are building a radio-controlled model airplane of a Boeing 737 at, say, 1/100 scale. It will likely take you 250 hours in your basement to build it from a kit before you can fly it at your local park. Now try scaling that up by comparing it to building one of Boeing's actual 737 aircraft. It probably takes many millions of hours to build one of those. More importantly, while you could build a model in your basement on a whim, Boeing needs to invest large amounts of money and take on enormous risks. Look at their 787 Dreamliner. Two fatal crashes, major software faults, employees claiming shoddy workmanship, and their entire fleet grounded. And this is from a company that has been building aircraft for over one hundred years, so they know what they are doing.

Therefore, carbon capture might be far more expensive and less effective when it scales up.

Another issue is that I am not crazy about storing CO_2 for eons any more than I am about storing radioactive waste. Think about how much will be stored after ten years; many millions of tons at least. Now think about what happens if something goes terribly wrong, say an earthquake or sabotage, that immediately releases all that stored CO_2 into the air at once. It's one thing to sip on a beer but quite another to chug a bottle of whiskey all at once.

What's the likelihood of that happening and what are the consequences? What happens when you store unheard-of quantities of CO_2 for decades or centuries even without an earthquake? Does it leak? Does it undergo chemical reactions changing its storage ability? No one knows.

"As an Aside, here's an unlikely problem but an interesting one. We know that adding CO_2 to the air increases the temperature, which is the whole point of removing it. But if adding CO_2 increases the temperature, then it must follow as the night the day that removing it will decrease the air's temperature. For

instance, we are now at over 400 ppm of CO_2 in our air. We would like to get that down to 280 ppm, the value before the Industrial Revolution took off. But if we went too far below this value, and as low as 190 ppm, we would bring on another ice age. Giant glaciers would move their way south into New York, Georgia, Montana, Florida—there would be no stopping them".

"Don't panic. Earth's substantial thermal inertia would put the brakes on this, and the heat trapped in the oceans would resist it. But eventually, if it continued, it would certainly cool things down".

"So here's the best part. Wouldn't this make for a great modern-day 007 movie plot? Imagine, an evil genius builds a gargantuan CO_2 vacuum cleaner and tries to drive the world into another ice age because he cornered the market on electric blankets. Brilliant".

Blocking the Sun—A Dangerous Game We May Be Forced to Play

Why not go to the source of the problem and block some of the sun's radiation from reaching the earth? As bizarre as it sounds, it is quite possible and it follows from what volcanoes have been doing for millennia.[vi]

Sulfate aerosols, when released into the stratosphere (66,000 feet up), reflect some of the sun's incoming radiation back into space, which counters the warming effect of global warming. To artificially create what volcanoes do, sulfuric acid is sprayed into the stratosphere, which then combines with water vapor to form sulfate aerosols less than a micrometer in diameter. Winds and temperature gradients disperse it over the globe.

Volcanoes provide experimental proof the idea will work. When Pinatubo erupted in the Philippines in 1991, it injected 20 million tons of sulfur dioxide into the air and the resulting aerosols reflected some of the sun's energy. Global temperatures dropped 1°F for several years [30]. That may not sound like much, but it is quite significant when compared to the temperature rise of 2.7°F (1.5°C) we are trying to avoid.

How do we do it? Keith calculated that we would need to inject 28,000 tons of sulfuric acid into the stratosphere each year in 2020 and increasing until we get to over one million tons by 2070 [31]. This will reduce Earth's warming trend by 1.5°F (0.75°C), a substantial victory. The cost was estimated to be $700 million per year.

It is not so easy to airlift 28,000 tons 66,000 feet in the air and Keith estimated

---

[vi] There are other geoengineering approaches possible, such as seeding the oceans with iron to increase their CO_2 uptake, but all of them have serious drawbacks.

that eleven or so high-flying jets would need to be used. A separate study stated that to develop these aircraft would cost over \$2 billion [32]. Not chump change and it will take time.

In its favor is the unmistakable fact that, unlike say carbon capture, sun blocking can be deployed with the technologies we now have in hand. No risky technological breakthroughs are required. It will work, of that we are sure, and we have some idea of its cost.

However, every silver lining has a cloud and this one is a doozy. Its unintended consequences might create a spate of new problems worse than the one we are solving.

But first, let me be clear and you must understand this. Sun blocking and other, similar climate engineering ideas do not solve global warming. It does nothing about the underlying problem of CO_2 accumulation which continues unabated. They do not reduce fossil fuel use, nor do they reduce any other greenhouse gas. They do nothing to slow down ocean acidification and coral bleaching. The global warming problem continues but now as a silent cancer growing without symptoms, which will kill us all the same.

The only thing sun blocking does is buy us time. Nothing more. Don't forget that. It stabilizes a dying patient, not cure him. It puts us on life support so we can survive long enough for other remedies, such as renewables and nuclear energy, to eliminate CO_2 from our air.

As CO_2 increases, its effect will get worse requiring ever more sun blocking sulfate aerosols. You can see the progression from Keith's numbers: 28,000 tons in 2020 jumping to over one million tons by 2070. That's a forty-fold increase or an almost doubling each year. How long do you think we can keep that up?

And we will need to do this every year without interruption. We will be a junkie looking for our next fix. What if we abruptly stop? There would be a sudden and large increase in temperature since the CO_2 that has been building up unnoticed is now no longer challenged by the sun blocking aerosols, and its influence goes unchecked.

And we are messing with a global experiment unlike any before. This is no Petri dish experiment where if something goes wrong you kill off a few specimens, no big deal. In this experiment, our planet is the Petri dish and we are the specimens; so yes, big deal. Rainfall patterns could change which could decimate local populations and the sulfate aerosols could damage the ozone layer. And more.

Sun blocking could also make the long-term problem worse if temperatures are artificially stabilized since without the reminder of hot weather, people may remain on their rumps satisfied with the world. That's human nature. Can you imagine how social media and political pundits would jump on this if global temperatures started to decline? Like sharks smelling blood in the water.

But sun blocking buys us time which is especially important since we are going to overshoot the 1.5°C limit by 2050. The problem is that once we overshoot, the CO_2 remains in the air forever, and no matter what we do to future emissions, even if we miraculously stop them completely, the overshoot remains. We have only two ways of getting out of this mess: carbon capture and sun blocking. There are no other possibilities.

Carbon capture pulls us back from the precipice. Sun blocking buys us the time needed.

But given the unknown ruin sun blocking could cause, we should take baby steps to learn as much about it before we go into any risky large-scale experiments. Modeling is a must. This will give us the theoretical underpinnings to guide the work and understand, not perfectly, but better than we do now, the landmines we could be stepping on. Next, lab experiments and then small field experiments need to be conducted before we go charging forth with anything more ambitious. Plus, a broad scientific and political consensus will be necessary.

I'm not a big fan of sun blocking, but I also realize we better start working on it now so we can put it in our tool box of solutions when the time comes to use it.

Decisions We Must Make

Both Conservatives and Liberals Must Be Convinced or Nothing We Do Matters

"*Get there the firstest with the* mostest."[vii] That's an old military slogan, pointing the way to win a war. Get to the battlefield first with overwhelming force. This strikes true with our efforts against global warming, which is a war like no other. We are not going to win it with just one faction of the U.S. population while ignoring the others. Imagine playing a football game and fielding only half your players. The outcome would be ugly. Worse still, imagine the half left on the bench charging onto the field to interfere with their own teammates. Hideous.

In the U.S. we either go with the full team, both conservatives and liberals,[viii] or we go home. Which means we need to accommodate, or at least not threaten, their other strongly held beliefs. Otherwise, we will alienate one or both groups into using global warming opposition as a proxy for what really bothers them.

[vii] From the 1917 *New York Times* and sometimes wrongly attributed to a Confederate general.

Basically, we need to uncouple the global warming issues from the other issues that separate conservatives and liberals so they each can focus on global warming. We don't want one faction sabotaging the global warming efforts of the other because they have been sidelined with their issues minimized and ignored.

So we need a consensus from both political spectrums on how we should proceed. Not easy, and compromise on both parts will either happen or, again, we go home. Let's say your solution to global warming is to rely entirely on free markets to decide and let the fossil energy industry run wild. Good-bye liberals. Or let's say you try to pass laws that mandate zero emissions by 2050 irrespective of the disruption to individuals' lives and liberties. Good-bye conservatives. Try either approach and we are left with a scorched earth policy instead. Literally.

An example of how not to do this is Joe Biden with his "I'm coming for you and I'm taking you down" message to the gun industry [33]. Seriously, Joe, do you think that will heal the partisan divide this country has gone through? Do you think you will have any credibility with the conservatives on global warming now?

What's required? Leadership. Now, there's a quaint idea that both Republican and Democratic politicians have forgotten about. We need political leaders who are able to rise above the internecine squabbles and put their own interests aside. What is best for either political party must be what is best for our country first, not the other way around.

Sounds simple. It's not. And you're not going to like what I write next. Our political leadership is exactly who we voted into office. You and me. The stork didn't bring them. We want our politicians to step up to be leaders? Then we need to step up and be thinkers who can see past the rubbish thrown at us by countless special interest groups that care less about us than about their next quarterly bonus. We must resist the influence of the large money donors who pretend they are on our side when they just want to fill their own pockets. We need to use our System 2 and start thinking critically and we need to devote the time it requires. Global warming will bite us badly and it demands attention.

And let's not try to pass the buck. Global warming exists because of people, not because of some abstract evil that we have no control over. The dire political leadership vacuum is because of us. We are responsible for our leaders' decisions as much as they are. We voted them into office, and we have the power to vote them out. Own it or we are doomed.

"A democracy gets the government it deserves." Let's deserve better.

As difficult as all this seems, it gets more complicated still. Both liberals and

---

[viii] For global warming purposes, I assume independents will join one camp or the other.

conservatives are in denial about global warming but in opposite directions. Liberals believe we can fix global warming as easily as replacing a flat tire without much expense or disruption. Conservatives don't believe global warming is happening and if it is, we are not responsible.

Both denials have a common root: an opposition stemming from inflexible Metabeliefs and an almost childlike resistance to any changes to their lifestyle and status quo. For some, how we live is as sacred as any religion, and maybe more so since it is concrete without the ambiguity of religious symbolism. Our lives are before us, plain as day and we never get away from it. It cannot be denied, so to accept global warming, or the unpleasant consequences of fighting it, is to suggest that we will need to revise everything we have spent a lifetime understanding and building. To accept it is to tell our children and grandchildren we have failed and we are giving them a lifelong burden of costs and social upheaval that they will need to deal with after we are gone.

On the other hand, to ignore it is to live happily ever after, even though in a fantasy that sooner or later will burst. Many are hoping later so they will not have to deal with it. It is much easier to deny then to act.

Again, it is not so much that conservatives don't believe global warming is happening or that liberals don't see that our lives are going to be profoundly changed. No, it's more that they each refuse to allow these thoughts to become part of their emotions and beliefs. Much better to lock them up in an intellectual prison that can be viewed from afar with no danger to their sense of security and self-worth.

There is no easy way out of this. It requires us to unleash our System 2, so we can undertake the critical thinking and judgments required to break out of our denials and assess what the dangers are and what we can do and what we must do. It's not an overnight transition but will require hard work, diligence, and some painful reckoning about ourselves and our beliefs. We need to adjust our Metabeliefs to accommodate what may be uncomfortable but necessary. As Benjamin Franklin said, "no pain no gain," and I suppose that means we are in for a lot of pain.

A Bridge Too Far?

The First Amendment's freedom of speech guarantee is one of the cornerstones of our democracy and should never be compromised. It gives all citizens the right to publicly state their opinions without risk of government retaliation. As Katherine Timpf put it, "[The First Amendment] protects a right our Founders considered crucial in maintaining a free society" [34]. And a 2019 poll showed that 63 percent of people would miss the freedom of speech if it were taken away [35].

The First Amendment protects the press from censorship, allowing them to force government transparency, so not much can be hidden, which acts as an additional check and balance on all three government branches. Contrast this with countries where the press is government controlled. North Korea, China, and Russia, to name a few. These countries are dictatorships in all but name with their citizens subjugated specifically because their press is controlled or nonexistent and individuals risk arrest if they say anything not sanctioned by the government. A sad example is Dr. Li, who first sounded the alarm about the coronavirus in China and was detained by the police and forced to sign a bogus confession. He was a hero and died from the virus while trying to save others.

So let's raise a toast to our First Amendment.

But freedom of speech is not freedom of cold cash, the ability to spend outrageous sums of money on members of Congress and advertising to achieve political goals that voters would never otherwise allow. Look what Political Action Committees (PACs) do. They contribute money directly to political campaigns and in 2018, raised $3.1 billion, of which $500 million went directly to candidates [36].

And there are 11,102 registered lobbyists in D.C. who spent $1.75 billion in 2019 trying to affect politicians' votes [37].

The problem with global warming solutions is that the fossil fuel industry lobbies spent almost $56 million in 2018 and pushed for "market-based solutions," which means they can do whatever they wish [38].

Don't for a minute think that money can't buy votes and decide elections as well as legislation, laws, rules, and everything else that controls how our country operates. Madison Avenue has been doing it for decades, telling us what to buy to be one of the beautiful people. It has nothing to do with facts but rather with how to manipulate human nature by pressing our hot buttons and relying on our System 1, overruling any critical thinking from System 2.

As the great sage George Carlin said,

"Advertising sells you things you don't need and can't afford, that are overpriced, and don't work. And they do it by exploiting your fears and insecurities, and if you don't have any, they would be glad to give you a few."

Money talks, and when used by skilled lobbyists who, along with their supporting cast, have had years and large budgets to research and understand psychology and human nature, it gives them the knowledge and experience to manipulate us.

Money also does more than just talk. It shouts, screams, yells, and bellows until it gets its way, which it almost always does. You and I are just corks bouncing on an ocean of influence with no choices or defenses. It is an assault that we simply cannot defend.

I am not alone. Sixty percent of Independents, 80 percent of Democrats, and 33 percent of Republicans believe our economic system is rigged against us for the benefit of those in power [39].

An example of lobbyists killing good legislation was the Fair Repair bills that would have required, quite sensibly, manufacturers to provide consumers with manuals, hardware, and software to allow them to repair their products "from phones to tractors" [40]. Opposition by Apple, General Motors, and John Deere, among others, killed these attempts.

Spending money is not free speech. Just the opposite. It is collusion disguised as free speech and should be greatly restricted to protect voters. Laws should be enacted that would curtail the lavish spending and influence peddling that accompanies it.

The Supreme Court disagrees. In their 2010 decision *Citizens United v. Federal Election Commission*[41], they ruled that any entity, be it corporations, nonprofits, labor unions, or others, has the right to spend money, and as much of it as they like, on political messaging.[ix] The Supreme Court's decision caused an avalanche of money pouring into political causes. Party hearty was their unmistakable message in a *1984*-esque fashion, green-lighting anyone with money to manipulate voters and control our country. And it spawned SuperPACs, legal entities that can spend ungodly amounts of money advocating or attacking any and all political candidates. In the 2019–2020 cycle, they collected over $3 billion [42].

Still not convinced? Just eleven wealthy individuals, six Democrats and five Republicans, donated $1 billion to SuperPACs over eight years [43]. So how much is $1 billion? Imagine someone at your front door offering you $1,000 to vote a certain way. Now imagine there is a second person behind them, ready to do the same and a third and so on. To reach $1 billion would require a line of such donors stretching from Washington, D.C. to Dallas, Texas. If that isn't outsized influence, especially when only eleven individuals are involved, then I haven't a clue what is.

In supporting their decision, the Supreme Court's majority argued that the government had no place in determining whether large expenditures distorted an audience's perception [44], and further, "because money is essential to disseminating speech . . . limiting a corporation's ability to spend money is unconstitutional" [45].

This misses the point by a country mile. Justice John Paul Stevens, in his dissenting opinion, stated, "Corporations make them dangerous to democratic elections . . . have perpetual life, the ability to amass large sums of money . . . that few individuals can match . . . limited liability . . . no morality, no purpose outside of profit-making, and no loyalty" [46].

One person, one vote? Bah. It doesn't exist anymore than unicorns do. It's more like one dollar, one vote, or better still, one million dollars, one vote.

Therefore, I advocate for a constitutional amendment to end all donations to congressional members and to abolish lobbying that pushes money into political causes. And let's make elections federally funded as other countries successfully do.

Further, I would ban all PACs, SuperPACs, anyone giving money to members of Congress and anyone spending large amounts of money to influence elections. Let congressional representatives earn their positions the old-fashioned way: by convincing voters they are deserving.

I realize I am tilting at windmills, and opposition will be fast and furious. The money will not go away. It's too late for that. The precedent has been set, and there is too much experience and self-interest for it to disappear. Lobbyists have a $1.75 billion annual business that provides for some very rich lifestyles they are not going to simply give up. They are going to fight it tooth and claw and will turn it into the mother of all political wars. But at least the discussion of a constitutional amendment shifts the battle from controlling lawmakers to convincing voters, which is going in the right direction.

I know there is more to it than just money. The voters do get it right sometimes and can see through the fog, and often find the right path. This is exactly what I am pushing for, but more frequently, with more certainty, and without a lot of fuss.

Put the power back in our hands and away from the powerful, as our Founders intended.

A sidebar to this is that our Electoral College is used to decide who becomes our next president. Terry Gross makes a compelling argument that modifying it to a popular vote instead would guarantee each voter has a say in the election, which is not the case now [47]. The way it currently works, many voters have no influence on the outcome because they are a minority in their state, and since most states use a

---

[ix] Though this decision does not allow for direct campaign contributions.

winner take all, their vote affects nothing. For instance, if you were a Yankee fan watching the 2004 American League playoffs at a Red Sox fan's house in Massachusetts, and the occupants were polled as to who will win, the Red Sox would be chosen no matter your vote or your knowledge of baseball. You are in the minority and your opinion and vote are completely immaterial to the poll's outcome.

Similarly for our presidential elections. In Texas, which always votes Republican, Democratic voters don't count and in California, which always votes Democratic, Republican voters don't count. They might as well stay home and knit since they have absolutely no influence on the election. Except for maybe twelve states, most states are like that. That is a vast number of voters with no say. As Gross points out, this then leads to candidates taking extreme positions since there is a very limited pool of voters they need to convince. Making it a popular vote instead, would likely drive candidates to a more moderate stand.

And don't get me started on gerrymandering, a disgraceful tactic used by both sides of the aisle. This is when the party in power moves the election boundaries within their state to favor their candidates. It's that Yankees fan being forcibly moved to Massachusetts. His vote disappears into a black nothingness. Making electoral boundaries fair and representative of its voters is another item on my bucket list. I better live a few hundred years if I am going to get to them all.

So in alignment with the theme of this section, the president should be decided by popular vote and not by Electoral College votes. One person, one vote.

The Fossil Fuel Energy Industry is the 800-Pound Gorilla. Are They Friends or Foe?

It would seem easy to blame the fossil fuel energy industry for our global warming woes. After all, they provide the heroin-like fuels we are addicted to that are choking our air. Fossil fuels account for most of the greenhouse gases emitted, and the worst of them, CO_2 lasts forever.

And they have consistently resisted global warming actions. Exxon shareholders rejected motions to combat global warming "arguing that too much remains unknown about the threat of climate change," [48] which is an identical tactic the tobacco industry used for many years to deny the significant health issues of smoking. Mirroring the tobacco industry's tactics is nothing to be proud of.

Further, in 2015, Exxon chairman Rex Tillerson stated, "The reality is there is no alternative energy source . . . to replace fossil fuels." He further made the tradeoff between "future climate events which could prove to be catastrophic but are unknown . . . and the more immediate needs of humanity today" [49].

That's a lot to unpack. At least he admits global warming would be catastrophic. But sorry, Rex, problems from global warming are not unknown. They are as well known as you know the back of your billfold. I agree, no energy source can replace fossil fuels, but that is no different from saying there are no fossil fuels that can replace renewables or nuclear. Each is different, with advantages and disadvantages.

It doesn't stop there. Exxon was accused of misleading investors about the cost of global warming to its bottom line. It told investors it was using higher costs of global warming than it actually was using in its own internal accounting [50], thus making it sound like its investments were less risky than they were. Peabody Energy, a coal giant, did the same and undervalued the risk it faced due to global warming [51].

The fossil fuel industry is also big. The ten largest fossil energy companies have a combined annual revenue of $2.75 trillion and employ over 3 million people [52]. There are nearly fifteen hundred oil and gas firms listed on stock exchanges with a combined worth of $4.65 trillion. Exxon alone is worth $425 billion [53]. That is staggering.

And they are powerful too. They spent almost $2 billion on lobbying for global warming legislation from 2000 to 2016, outspending environmental and renewable energy groups by ten to one [54]. The five biggest publicly traded companies, BP, Shell, ExxonMobil, Chevron, and Total Global, spend $200 million per year on lobbying efforts to block or delay global warming actions [55].

But let's take a deep breath and consider their side of the story. Corporations, fossil fuel or not, were created to maximize profit, which is what stockholders rightfully demand. To ask the shareholders to surrender some of that value for global warming actions is to unfairly tax them.

Further, they have a reason to resist as they have a lot to lose from global warming actions. The International Energy Agency projected that fossil fuel demand could drop by 22 percent by 2040 [56]. And if you believe the IPCC, then fossil fuels will need to be all but eliminated by 2050. This could wipe out $0.75 trillion of fossil fuel companies' value [57]. Think about that. Let's say someone stole your identity and took everything you owned—bank accounts, car, house, investments, salary. Is that something you would roll over for? No. Neither will the oil companies nor their 3 million employees and countless stockholders.

You can see what we are up against. We are dealing with enormous, powerful, almost unstoppable forces. Do we go to war against them? Given what they provide, that would be like going to war with our own kidneys. The fossil fuel industry helped bring us out of the colonial era, which had a life expectancy of thirty-five years and into the modern age of jumbo jets, medicine, hospitals, smartphones, and the highest standard of living in the history of our planet. No complaints from me.

No, not war, but an alliance, which is far more useful. Let's make them part of the solution, not the scapegoat we seem to be always looking for. Fortunately, there is common ground that would both benefit the fossil energy industry and aid everyone in our fight against global warming.

Carbon capture nicely fits into this common ground since the more carbon is captured, the less malignant burning fossil fuels will be. One approach is to ramp up carbon capture and reduce fossil fuel use, as far as is practical, such that the CO_2 being released by the remaining fuels is captured, resulting in zero net CO_2 emissions.

This actually is the best of both worlds since the carbon capture soaks up CO_2, but fossil fuels are still used in those sectors that must have the extraordinary energy density they contain. Commercial aircraft are not likely to switch to electric engines and certainly, the military will not. They need the large energy packed into a small volume that fossil fuels offer.

But there is another, more fundamental issue: the economics of supply and demand. As demand for fossil fuels dries up, all else being equal, fossil fuel prices will come down. Therefore, absent draconian laws prohibiting fossil fuel use, some businesses will still use it as the most cost-effective approach, even with carbon taxes included. Hence, fossil fuel use is not going to disappear altogether, which means CO_2 capture is necessary, so fossil fuel companies might as well get into the game, especially as it fits in nicely with their main product.

Fossil fuel energy companies dabbled in renewables for a time. BP ended its solar business and Chevron sold its renewable energy division [58]. Now, five years later, BP is committed to ending its own greenhouse gas emissions with renewables by 2050 [59]. Hopefully, that also means they will get back into the renewable business.

This pivot makes a lot of sense for them. First, fossil fuel burning will go down, hurting their business. Second, fossil fuels are also used in making plastics, such as plastic bottles, straws, and packaging, and there has long been a push to phase these out. China plans to ban plastic packaging by 2025 and other countries might follow [60]. The European Union plans to ban plastic drinking straws, plates, and cotton swab sticks. Canada plans to ban all single-use plastic items.

So it is in everyone's interest if the fossil energy companies invest in renewable technologies with an eye toward turning them into profitable businesses that can replace fossil energy business, which will inevitably fade.

Turning to carbon taxes, my instinct is that the fossil fuel energy companies would oppose them. I was wrong. They have stated they are in favor of a carbon tax but only if it replaces emissions regulations and if nuclear plant subsidies are ended [61]. This seems like a Trojan Horse to me. Unlimited emissions and kill nuclear in return for a carbon tax? Thanks, but no thanks. Still, it's interesting in that it signals the oil industry is moving from outright stonewalling to encouraging statements, no matter how hollow. That's progress of some sort, I suppose.

One key is to convince oil company stockholders that global warming is a "material risk to their core business" [62]. Now we're talking. If they could internalize that, then they would be willing to plan for a transition to other forms of energy rather than fighting it with smoke and mirrors and might. This would naturally lead to a search for other sectors that could provide profits to the fossil fuel companies.

And taxpayers could grease the way by providing incentives for their entry and sustaining work into the carbon capture and renewables industries. The fossil fuel energy companies get a windfall of public relations and they protect their shareholders by moving away from fossil fuels without the risk of losing a great deal of value.

That's our part of the bargain and they need to realize nothing in life is free. Their part is to vigorously pursue renewables and carbon capture by developing the technologies and supporting others to do so. They must devise a credible business plan that will take them across the chasm from fossil energy as a major source of sales to other ones, even if it means a reduction in overall market share and annual sales. And they must cease lobbying against nuclear energy and must support a carbon tax.

I don't mean to be overly prescriptive, as we all need to be flexible and willing to change our targets and resources as we learn more and figure out what works and what doesn't. But this is a start.

Remember, fossil energy companies shouldn't be blamed for their past sins, as no one really appreciated how serious global warming problems were going to be.

But in the same vein, they cannot be absolved of future sins either. Besides, they are not the only villain in this story. There are two others. One of them is reading this book and the other wrote it. We are the ones eagerly using their products. So let's work together. Not apart.

Fool's Gold

It is important we reduce our carbon footprint. We need to go from beef, to chicken, to fish, to vegetables and fruits. We need to drive less and with more fuel-efficient cars. Industry must reduce its energy use and become more productive. There is much we can and should do.

But all these actions solve nothing by themselves. They are too little too late, although what they can do is lighten the load a bit for the heavy lifters from the previous sections. It eases their burden so they have a better shot at succeeding.

The problem though, is with the fossil fuel companies. Remember them? They are the ones, as Stevens wrote above, that "have perpetual life, the ability to amass large sums of money . . . that few individuals can match . . . limited liability . . . no morality, no purpose outside of profit-making, and no loyalty." They will jump on this like a Three-card Monte dealer with newly arrived gullible tourists. They will claim that this is all that's needed which will get them off the hook.

It's the sugar companies all over again, who told us exercise will offset a bad diet. It doesn't. A bad diet is bad. Period. The same with our fossil fuel friends. They will argue that as long as we are screwing in LED light bulbs, all is well and they can go about their business relieved of their responsibility to correct the world-changing problems they are intimately connected to.

Imogen West-Knights calls it "greenwashing" [63] I call it brainwashing. We must not fall for it. Fossil fuel companies must be held accountable and no matter the noble actions of individuals and companies in slowing global warming, fossil fuel companies are still the tip of the spear pointing at us. And we should never allow them to spin their way out of their responsibility.

So we shouldn't be fooled by the energy companies any more than the sugar companies should fool us. Or the Three-card Monte dealers.

But Ultimately, It Is Up To Us

Four Things You and I Must Do

1. Global warming must become an intrinsic part of our Metabeliefs. It must be as

intuitive as knowing how we take our coffee in the morning and what our evening bedtime routine is. We need to understand and accept global warming emotionally as well as intellectually. Nothing less will do.

2. We must become adept at seeing through the lies and conspiracy theories that appear to be everywhere these days. We need to resist the allure of appealing explanations that are as phony as the Nigerian emails offering us millions. We can only see the truth if we are armored against lies.

3. We must pick leaders who will take action against global warming. Politicians certainly, but also the leaders of the many groups we belong to. People follow their leaders. This is a simple fact. So let's be sure we follow honest, knowledgeable leaders who see global warming for what it is and not for what charlatans want them to see.

4. Remember; the key to global warming is burning fossil fuels. Memorize this last sentence. So anything we can do to reduce our fossil fuel use is critical. Reduce it or, better still, eliminate it. No, that doesn't make global warming go away. Life is not that simple. But the more we reduce, the easier we make it for the heavy lifters in the previous sections to get the job done.

Connecting The Dots

We need two parallel approaches, and both are essential. Either one by itself will not work.

The First is to Deploy Equipment to Directly Reduce or Remove CO_2

1. Renewables

- Solar and wind are essential to our campaign against global warming as they will provide unlimited energy without any greenhouse gas emissions. They are well-developed and economical.
- Hydrogen and recycling should be considered as renewables.

2. A carbon tax

- This is a fee added to all fuels based on the amount of carbon they contain and therefore, the amount of CO_2 they will emit.
- It levels the playing field of all carbon and non–carbon-emitting fuels and will reduce the demand for fossil fuels.
- Cap and trade are similar to a carbon tax but instead rely on having the marketplace decide on the value of the carbon offsets.

3. Nuclear fission

- Nuclear fission energy produces large amounts of energy without any CO_2 emissions and is well-developed. There are no technical issues so we should be using more of it.
- Nuclear fission does, however leave a trail of destructive nuclear waste that we will need to deal with for many thousands of years.

4. Nuclear fusion

- Nuclear fusion is the new kid on the block. Well, actually the old man on the block who never matured.
- Like fission it produces carbon free energy.
- But it also does so with no leftover nuclear waste.
- Plus, it uses cheap seawater as a fuel.
- Unfortunately, fusion energy has never produced net energy and has a very long way to go to be widely commercially deployed.
- Nonetheless, fission and fusion serve the greater good. Fission nuclear energy is used now, and once fusion is fully developed, it can phase fission energy out.

5. Natural gas leaks and cow burps

- These also produce greenhouse gases.
- They are easy to address. Plug the leaks and eat less beef.
- Easy to say anyway, but it will take money and willpower, which are always in short supply.

6. Carbon capture

- It removes CO_2 already in the air.
- There is no other way of removing it.
- Without carbon capture there would be little hope of ever turning the clock back. With it, there is.

7. Blocking the sun

- This is done by injecting sulfates in the air.
- It will buy us time but it is a band-aid at best and a very dangerous experiment at worst.

- Still, it should be pursued so we can better understand it if we are forced to use it.

But what good are the above if we don't have the leadership to back it up?

8. We need all Americans, not just a few.

- Conservatives, liberals, and all other political and religious stripes must be on board fighting the same fight or we will fail.
- It's all for one or everyone for nothing.

9. Vast amounts of special-interest money control our country.

- We must oppose the moneyed interests that care only about their next yacht and nothing about global warming and us. Money must be removed from influencing decisions and laws.
- This may require a constitutional amendment, but the cost and effort are worth it if we can go back to those innocent days of one person, one vote and no longer need to worry about those with money and power unfairly ruling our lives.

10. Is the Electoral College fair?

- Maybe as far as it goes, but it marginalizes so many of us whose votes mean nothing because our state's Electoral Votes will be decided without us. Eliminate it so we again can get to truly one person, one vote.

11. Fossil fuel companies

- Energy companies, such as Exxon, have been selling us fuels for over a century and have served us well.
- But now their byproduct of CO_2 is coming to collect its long overdue bill.
- These very same fossil fuel companies are fighting against the necessary changes we desperately need to avoid global warming.
- Fossil fuel companies are large and powerful.
- It would be best if we work with them on common goals.
- Carbon capture and renewables could fit into their corporate plans. We need to encourage it.
- Their shareholders must internalize the risks they face from riding the fossil energy horse for too long.
- Taxpayers should make it easier for energy companies to transition to carbon-free energy.

- In return, energy companies need to embrace a carbon-free world.
- But let's not be fooled either. Doing our part does not relieve the energy companies of their responsibilities. "Greenwashing" must not be allowed.

And never forget, *it is up to us*. After all, we are all we've got.

Afterthought: On This Day in History, 124 Years Ago

On this day, more or less, 124 years ago, Svante Arrhenius received the Nobel Prize for discovering CO_2 is a greenhouse gas. At his acceptance speech, he stated, "On this day, more or less, 124 years from now, society will avoid all global warming effects thanks to my discovery since they will certainly take all the necessary precautions early enough to make it easy and effective. I can now go to my grave in peace."

Prompting modern-day historians to conclude "this guy may have known his chemistry but for God's sake keep him away from crystal balls." I agree. We should leave that to the experts, the psychics.

Svante Arrhenius really did win the Nobel Prize for discovering CO_2 is a greenhouse gas in 1896. The rest I made up.

However, he did suggest (if you believe anything I now write) we burn more coal to warm up the planet which in hindsight sounds as ludicrous as his future prediction, but he did indeed say it. Trust me. Which anticipated the now oft-quoted warning be careful what you wish for, so may be he could predict the future after all. Well, not exactly since that expression comes from *Aesop's Fables* a lot more than 124 years ago, more or less. This is starting to feel like an Escher painting.

Which finally sets me up to use one of my favorite cinematic lines:

"*You knew then! And you did nothing*"

—Julius Levinson, *Independence Day*

Yes, we knew back in 1896 all about greenhouse gases, and yes again, we are doing nothing or if not nothing, then not nearly enough, which for all practical purposes is nothing.

Somehow I doubt Svante Arrhenius is resting as peacefully as he predicted.

References

[1]   Special Report on Global Warming of 1.5oC, Summary for Policymakers, Intergovernmental Panel on Climate Change, 6 October 2018.

[2]   Electric Vehicles Sales: Facts & Figures, EEI, April 2019, https://wwworocobrecom/wp-content/uploads/2019/12/FINAL_EV_Sales_Update_April2019pdf 2019.

[3]   Weise E. Tech Firms Like Google, Amazon Push Power Companies toward Solar and Wind, a Blow to Coal," USA Today, April 2018.

[4]   Dudley D. Renewable Energy Will Be Consistently Cheaper Than Fossil Fuels by 2020, Claims New Report," Forbes, 13 January 2018.

[5]   Page S. The Energy Revolution Is Actually Happening Right Now," Think Progress, 4 May 2016; "Cost and Performance of Generating New Technologies, Annual Energy Outlook 2020," EIA, January 2020. 2020.

[6]   Fairly P. H_2 Solution. Scientific American February 2020.

[7]   Mosher D. Here's What Earth Might Look Like in 100 Years—If We're Lucky, Business Insider, 14 November 2018.

[8]   Chandler DL. "Larry Linden: Big Changes Needed to Avert Possible Climate 'Catastrophe,'" *MIT News*, 15 January 2015; Metcalf, G. E., "On the Economics of a Carbon Tax for the United States,". Brookings Pap Econ Act 2019.

[9]   Metcalf GE. On the Economics of a Carbon Tax for the United States. Brookings Pap Econ Act 2019; 2019(1): 405-84.
        [http://dx.doi.org/10.1353/eca.2019.0005]

[10]   Scientists Issue Carbon Price Call to Curb Climate Change," Phys Org, 10 July 2015.

[11]   How Much Electricity Does a Nuclear Power Plant Generate?," US Energy Information Administration, https://wwweiagov/tools/faqs/faqphp?id=104&t=3

[12]   Markandya A, Wilkinson P. Electricity generation and health. Lancet 2007; 370(9591): 979-90.
        [http://dx.doi.org/10.1016/S0140-6736(07)61253-7] [PMID: 17876910]

[13]   Rettner R. More Than 250,000 People May Die Each Year Due to Climate Change," Live Science, 7 January 2019.

[14]   Roberts D. How to Save the Failing Nuclear Power Plants That Generate Half of America's Clean Energy 2018.

[15]   Cho A. The xxx. Science 2019; 363(6429): 22.

[16]   Roberts D. How to Save the Failing Nuclear Power Plants That Generate Half of America's Clean Energy. 2018.

[17]   Phillips L. The New, Safer Nuclear Reactors That Might Help Stop. Clim Change 2019; 27.

[18]   Ibid.

[19]   Fueling the Fusion Reaction, ITER.

[20]   Nature's Perfect Energy Source, Fusion Industry Association.

[21]   Cost of Nuclear Fusion.

[22]   Krupp F. These Technology Trends Can Pave Way for Global Climate Action.

[23]   Spence S. Probing Question: Are Cow Burps Contributing to Global Warming? 2011.

[24]   Hansen J, Kharecha P. Cost of Carbon Capture: Can Young People Bear the Burden? Joule 2018; 2(8): 1405-7.
        [http://dx.doi.org/10.1016/j.joule.2018.07.035]

[25]   Freeman M, Yellen D. Capture That Carbon. Sci Am 2018; 319(2): 11.
[http://dx.doi.org/10.1038/scientificamerican0818-11] [PMID: 30020909]

[26]   Service RF. Cost of carbon capture drops, but does anyone want it? Science 2016; 354(6318): 1362-3.
[http://dx.doi.org/10.1126/science.354.6318.1362] [PMID: 27980164]

[27]   Conniff R. Scrubbing Carbon from the Sky. Sci Am 2019.

[28]   Berardelli J. Fighting Climate Change Would Boost Economy, Study Finds 2020.

[29]   Newell R, *et al.* Global Energy Outlook 2019: The Next Generation of Energy 2019.

[30]   Fountain J. The Next Big Volcano Could Briefly Cool Earth, NASA Wants to Be Ready 2018.

[31]   Rotman D. Intentionally Engineering Earth's Atmosphere to Offset Rising Temperatures Could Be Far More Doable Than You Imagine, Says David Keith But Is It a Good Idea?

[32]   Berardelli J. Controversial Spraying Method Aims to Curb Global Warming 2018.

[33]   Keane L. Joe Biden's Plan to Shut Down the Firearms Industry. Natl Rev 2020; 17.

[34]   Timpf K. Trump Should Stop Suing the Press. Natl Rev 2020; 10.

[35]   Bote J. 92 percent of Americans Think Their Basic Rights Are Being Threatened, New Poll Shows 2019.

[36]   Political Action Committees," OpenSecrets.org.

[37]   "Lobbying Data Summary," OpenSecrets.org.

[38]   Hillstrom K E, Arke R. Fossil Fuel Companies Lobby Congress on Their Own Solutions to Curb Climate Change 2019.

[39]   Leonard C. Yes, America Is Rigged. Here Is What I Learned from Reporting on Koch Industries. Time 2019; 26.

[40]   Hernandez K, Rebala P. Puppies, Phones, and Porn: How Model Legislation Affects Consumers' Lives 2019.

[41]   "Citizens United v. FEC," *Wikipedia.*

[42]   "Super Pacs," OpenSecrets.org.

[43]   Lee M Y H. Eleven Donors Have Plowed $1 Billion into Super PACs since They Were Created 2018.

[44]   Hernandez K, Rebala P. Puppies, Phones, and Porn: How Model Legislation Affects Consumers' Lives 2019.

[45]   Ibid., p. 7.

[46]   Ibid., p. 11.

[47]   Gross T. Electoral College 'Not Carved in Stone': Author Advocates Rethinking How We Vote 2020.

[48]   Zarroli J. Exxon Mobil, Chevron Shareholders Reject Resolutions Aimed at Battling Climate Change 2016.

[49]   Ibid.

[50]   Exxon Accused of Misleading Investors on Climate Change 2019.

[51]   Zarroli J. Coal Giant Peabody Accused of Misleading Investors about Climate Change 2015.

[52]   "List of Largest Companies by Revenue," *Wikipedia.*

[53]   Casella C. When the Fossil Fuel Industry Pops, It Will Be Way Bigger Than the 2008 Financial Crisis 2018.

[54]   Fossil Fuel Interests Have Outspent Environmental Advocates 10:1 on Climate Lobbying. Yale

Environmental 2018; 360: 19.

[55]   McCarthy N. Oil and Gas Giants Spend Millions Lobbying to Block Climate Change Policies. Forbes 2019; 25.

[56]   Krauss C, Schwartz J. Exxon Investors Seek Assurances as Climate Shifts, Along with Attitudes 2016.

[57]   Bousso R. Tighter Climate Policies Could Erase $23 Trillion I Companies Value Report 2019.

[58]   Zeller T. The End of the Partisan Divide over Climate Change. Forbes 2015; 18.

[59]   Holmes F. The Case for Pivoting into Renewable Energy 2020.

[60]   Ibid.

[61]   German B, Harder Am. Industries Latest Climate Shift. Axios 2019.

[62]   Domonoske C. Shareholders Push Exxon to Disclose Business Impact of Fighting Climate Change 2017.

[63]   West-Knights I. The 12 Arguments Every Climate Denier Uses—and How to Debunk Them 2020.

CHAPTER 13

Beyond a Reasonable Doubt

Ladies and gentlemen of the jury:

The fate of our planet rests in your hands. Take a deep breath and let that sink in for a minute. It's a heavy burden, I know. Global warming's unchecked rampage has already killed many thousands and has caused billions of dollars in damage. And it is just getting warmed up, so to speak. Your job is to decide whether its assault is real and, more importantly, what's to be done about it.

Let's review the evidence.

First, the murder weapon. Greenhouse gases, mostly CO_2, have been spewing from the fuels burned in homes, autos, factories, and airplanes. CO_2 has relentlessly altered our air, disrupting the beautiful balance between our planet and the life-giving sun that warms us. It has caused more heat to hit our planet than the planet is capable of removing. What happens when you raise your thermostat setting? The temperature goes up. It gets hotter. And so it has with our world.

This has produced giant storms, melting glaciers, raising sea levels, and extinctions seen only five times in the past 300 million years.

Next, the motive. We have been on an energy-rich diet since the Industrial Revolution, and it has propelled our society to a lifestyle never experienced, maybe never dreamed of, by our ancestors. All in all, it has done well by us. But it has gone too far in its insatiable drive for wealth, discarding our well-being for the addictive need for more wealth and still more after that. Now it is sacrificing our planet, us, and our children for that unreachable itch.

Finally, there was ample opportunity. Since the 1700s, we have been putting CO_2 into the air in ever-increasing amounts. Burning fuels was slow at first, but now it's more like a ruptured fire hose, and CO_2 has gone from 280 ppm to over 400 ppm in a blink of an eye. Never in the history of civilization, and 3 million years before then, have we put so much evil into our breathable air so quickly.

I've shown you the co-conspirators. The energy company shape shifters that provide the fossil fuels while hiding their true character—one of greed and power. Just like the tobacco companies before them, they have expertly distorted the truth, blinding us from seeing the cliff we are rushing toward.

And let's not forget the enablers; the deniers and denizens of social media, who are eager to betray us for a few pieces of silver and website clicks. Those with large followings who will readily sacrifice our children's future for their padded bank accounts, fast cars, and large houses.

You have seen how they weaponized lies to stop us from seeing the painful facts of global warming. To them, truth's destruction is acceptable collateral damage. In fact, it is preferable since it is a game of control as much as anything. Control us, so we don't swallow the red pill and see them as they truly are.

I've shown you how to get past their bluster, the smoke and mirrors, the outright lies, the street muggings of our reason. You are well-armed to judge the truth. It's not difficult. It just takes a few tools that you now possess and some critical thinking.

When you consider all of this, ask yourselves a simple question. How can nearly every climate scientist be convinced of global warming's harm if it is not so? Let the answer guide your deliberations.

I know you will agree the evidence is overwhelming and cannot be denied by any rational person.

I now ask you to return a guilty verdict of crimes against humanity. Guilty of polluting our air and wrecking our home planet. Doing it willfully with intent and malice. And causing harm to all of us.

But I also ask you to temper your verdict with compassion and mercy. Sure, it has been known as far back as 1896 that CO_2 is a greenhouse gas and as deadly as a coiled viper ready to strike. But how could anyone have guessed the serious threat it was? We were on a giddy rocket ascent of wealth and advancement with no time to consider the consequences.

But even compassion and mercy have limits. The guilty must reform and must follow a path clearly laid out, difficult perhaps, but not impossible. We can solve global warming but only if everyone agrees to do so.

You must judge the guilt and what should be done. But remember, you are not just the jurors. You are also the victims. And the accused.

Choose wisely.

The Prosecution rests.

EPILOGUE

Albert Calabrese jumped into the Ardennes Forest with the 82nd Airborne in 1944. At 23, he was tough and brutal though a bit naïve. As soon as he shed his harness, an enemy soldier charged at him. Thinking it was a joke, he smiled and put up his hand in mock defense. As I said, Calabrese was naïve. His hand deflected an onrushing bayonet, saving his life.

Calabrese fought his way through Europe earning a Bronze Star, two Purple Hearts, battle ribbons, and commemorative cords from Belgium and Norway.

And while the 82nd Airborne were special soldiers, maybe the most special; in many ways, Calabrese was no different than the millions of other men and women in the war and at home who shaped the U.S. into an economic and political powerhouse, the likes of which had never been seen before. As Shakespeare might have said, they may not have been born great, but they achieved greatness whether it was thrust upon them or not. And 75 years later, we are still reaping the rewards of their sacrifices and accomplishments.

Now it is our turn. We are in a similar position where global warming problems have been thrust upon us and threaten our well-being, indeed our very existence. Storms, rising sea levels, famine, violence, and intolerable heat are all attacking us on multiple fronts. And once again, we are in a fight for our lives.

What we do now will affect our nation and our world more so than at any time in humanity's brief existence. But look. Calabrese did it and he was just a green kid coming off the streets of Brooklyn. He was nothing special. At least not until he stepped through that door and hurtled toward the ground. That one step took him from an unremarkable placid life and into the history books.

I realize the responsibility we are now burdened with is causing us much angst and fear. But that load also brings an opportunity, few in history are fortunate enough to be a part of it. Remember, saving ourselves, our children, and all the generations coming after us will forever be celebrated with admiration and relief. Relief that the coming generations were spared from the worse of global warming, allowing them to accomplish unimaginable breakthroughs that will push humanity to newer and higher plateaus. And we will be the reason they could.

Just as Calabrese and his brethren knew they would emerge victorious, so will us.

Of this I am certain.

It is now our time to step through the door.

ACKNOWLEDGEMENTS

I wish to thank the following for their help and encouragement:

Diran Apelian
Albert Calabrese Jr.
Albert Calabrese Sr.
Paul De Saro
Ralph LaSardo
Scott Mellen
Antonella Pompo
Carlos Romero
Rob Stowe
Peter Stupak
Georjean Trinkle
Amy Weeder

SUBJECT INDEX

need for a court order if at any point you breach any terms of this License Agreement. In no event will any delay or failure by Bentham Science Publishers in enforcing your compliance with this License Agreement constitute a waiver of any of its rights.

3. You acknowledge that you have read this License Agreement, and agree to be bound by its terms and conditions. To the extent that any other terms and conditions presented on any website of Bentham Science Publishers conflict with, or are inconsistent with, the terms and conditions set out in this License Agreement, you acknowledge that the terms and conditions set out in this License Agreement shall prevail.

Bentham Science Publishers Ltd.
Executive Suite Y - 2
PO Box 7917, Saif Zone
Sharjah, U.A.E.
Email: subscriptions@benthamscience.net

BENTHAM
SCIENCE

Made in the USA
Columbia, SC
29 April 2023

3a101cb7-8b09-4a14-898d-1119a3781af6R02